ENGINEERING HITLER'S DOWNFALL

the Brains that enabled Victory

Lieutenant Commander Gwilym Roberts
CBE FREng FICE FIMechE RNR

Whittles Publishing

Published by
Whittles Publishing,
Dunbeath,
Caithness KW6 6EG,
Scotland, UK

www.whittlespublishing.com

ISBN 978-184995-386-3

Printed by CPI Group (UK) Ltd, Croydon CR0 4YY

The Author

Lt Cdr Gwilym Roberts CBE FREng FICE FIMechE RNR

Born in 1925, Gwilym Roberts was a teenager throughout the six years of the Second World War and thus well remembers its disasters and triumphs. Born in North Wales, he was raised and had his early education in Liverpool before going to the University of Cambridge in 1943 to study engineering. Two years later he joined the Royal Navy's Engineering Branch and saw service in Devonport and HMS *Sheffield.* He subsequently served part time with the Royal Naval Reserve.

Following demobilisation in 1947, he joined water engineering consultants John Taylor & Sons, with whom he worked on major projects in the UK and throughout the Middle East. After becoming senior partner he oversaw the merger in 1987 of the firm with transportation consultants Freeman Fox & Partners to form the Acer Group, of which he was the founder chairman. (Acer was later renamed Hyder Consulting and is now part of Arcadis.)

He served as president of the Institution of Civil Engineers in 1986–87 and of the Smeatonian Society of Civil Engineers in 2009. He has also been chairman of the British Geological Survey, the Football Stadia Advisory Design Council (established following the Hillsborough disaster), and a committee involved with the construction of the Second Severn Crossing. In January 2004 *The Sunday Times* magazine named him as one of Britain's four leading international engineers of the twentieth century.

He is passionate about the importance of engineering to national and global wellbeing and believes that the Second World War could well have been lost but for the timely development by scientists and engineers of the novel systems, machines, and weapons which were used by, and supplemented the valour of, the fighting forces.

Awards for technical papers he wrote when still in practice include the George Stephenson Medal and the Halcrow Prize from the Institution of Civil Engineers and the Silver Medal from the Institution of Public Health Engineers. Some of his more recent publications include:

- *Built by Oil* (Ithica Press, 1996) (Middle Eastern Post-war Projects)
- *Chelsea to Cairo* (Thomas Telford, 2006) (History of John Taylor and Sons)
- *From Kendal's Coffee House to Great George Street* (Thomas Telford, 1995) (ICE HQ buildings)

And the following papers (all by Thomas Telford for the Institution of Civil Engineers):

- 'Bridging in the Second World War: an Imperative to Victory' (with D. L. G. Begbie)
- 'F E Cooper (1841–1933): the Supreme Resident Engineer'
- 'How a Diver saved Winchester Cathedral'
- 'Middle East Archaeology'
- 'Middle East Postwar Engineering Projects'
- 'St Pancras Station – Victorian Cathedral of the Railways'

This book is dedicated to all the engineers and other technologists who participated in the Second World War, in recognition of their achievements and sacrifices and of the debt we owe to all those who made victory possible, thereby allowing us our present liberties.

Contents

Acknowledgements and Sources

First of all I should like to thank Admiral the Right Honourable Lord West of Spithead GCB DSC PC for very kindly writing the Foreword and for his complimentary comments about the book. With his experience commanding a ship in the Falklands campaign and as a former First Sea Lord and Minister for Security his observations carry considerable weight and are a further acknowledgement of the contribution to victory made by the wartime engineers and scientists and of the debt we all owe to them as well as to those in the armed services.

Since I was first introduced to him some three years ago, my publisher, Dr Keith Whittles, has been a fount of knowledge and advice and I am particularly grateful to him and to Rachel Oliver, Kerrie Moncur, Sue Steven and their colleagues for their advice, assistance and encouragement.

I am also particularly grateful to the principal officers and staff of the two professional engineering institutions that I have been most closely involved with and who have given me considerable encouragement and support. They are Dame Ann Dowling OM FRS FREng, Dr Hayaatun Sillem and Philip Greenish CBE, (respectively President, Chief Executive and the recently retired Chief Executive of the Royal Academy of Engineering) and Lord Mair CBE FRS FREng, Dr Gordon Masterton OBE FREng FRSE, Nick Baveystock FInstRE, Carol Morgan and Debra Francis (respectively the Institution of Civil Engineers' President, Chairman Panel for Historical Engineering Works and Past President, Director General, Archivist and Librarian). Others who have also given me considerable assistance include David Arscott of Pomegranate Press who has redrafted sections of the text to make them more readable, Bob Davis who has assisted with identifying the sources and copyright owners of the illustrations, Emily Dean of the Imperial War Museum for assistance and advice regarding the illustrations in the Museum's archives, Dr Phil Judkins (Trustee of the Purbeck Radar Museum Trust and formerly of the University of Buckingham) for advice and my daughter, Annabel Roberts, for coordinating recent inputs and assisting with the final editing of the book.

Others who have commented on draft texts, given specific advice or made valuable suggestions have included: Rear Admiral Ian Tibbitt (Trustee, Fly Navy Heritage Trust); Major-General K J (DZ) Drewienkiewicz CB CMG, formerly

Engineer-in-Chief (Army), the late Brigadier D L G Begbie OBE MC, Colonel Jim Joiner RE, Major R J Henderson REME (Director, REME Museum), Martin Skipworth (Researcher, Royal Signals Museum), Peter Elliott (formerly Head of Archives, RAF Museum and Chairman, Royal Aeronautical Society Historical Group), Dr Michael Purshouse FREng, (formerly Thales UK and BAE Systems), Lieutenant Commander Richard Olsen RNR, Donald Green (retired Fellow, Sidney Sussex College), Nicholas Rogers (Archivist, Sidney Sussex College), Michael Chrimes MBE (formerly of the Institution of Civil Engineers), Adrian Clement (Institution of Mechanical Engineers), Edward Kemp (Institution of Engineering and Technology), Rob Thomas (Institution of Structural Engineers), Richard Wakeford (British Library, Science Reference Enquiries), Wing Commander Tony Willenbruch RAF (formerly Clerk, Worshipful Company of Engineers), Zoe Edwards, (East Sussex County Council Libraries) and my son Matthew Roberts FCA. In addition to those mentioned individually above, many more have made valuable suggestions or comments or given me new leads or ideas to follow. I am greatly indebted to them all.

My principal sources of information have been the many books and articles that have been written about various aspects of the Second World War, in particular those which are listed in the Bibliography. On occasion such information has been supplemented by my personal recollections or those of colleagues. With so many books and articles written, it has not been practicable to identify the various information sources but practically all are in the books mentioned in the Bibliography. The biographical notes have mainly been prepared from information in the Oxford Dictionary of National Biography, other biographies, obituaries and Wikipedia.

I am most grateful to the copyright owners of the illustrations which have been reproduced and who are detailed on the next page. In particular I would like to thank the Institution of Civil Engineers (ICE) Library and Archives, for their permission to reproduce images from their archives.

Finally I should like to thank my wife Wendy for her unfailing support and encouragement without which it would not have been possible to have researched and written this book.

Illustration Sources

While every effort has been made to ascertain the sources of all illustrations, it is possible there may have been a few inadvertent errors and omissions – and in any such cases it is hoped that copyright holders will accept the apologies of the author and publishers.

Grateful appreciation is made to the following organisations and individuals for permission to reproduce the illustrations attributed to them, in addition to several Wikipedia images and other images in the public domain. All other images are my own or are out of copyright:

Amberley Publishing; Henry Barber – Aberdeen University Review; Neil Brown; Richard Brown; Stephen Budiansky; Stephen Canning; Chatham Historic Dockyard; Jane Fanshawe; Greg Goebel; Tony Hisgett; Imperial War Museum; Institution of Civil Engineers; Institution of Royal Engineers; Lidell Hart Centre for Military Archives; London Metropolitan Archive; Peter Mallows; National Archives; David Owen; Mark Postlethwaite; Gordon Rankine; Royal Society; Paul Russon; Robert Taylor, www.taylor-photo.co.uk; The Tank Museum; Tina Scott; family of Peter Verey; Tony Watson; www.purbeckradar.org.

Timeline for World War 2

Year	Date	Event
1	Mar-16	Germany annexes Czechoslovakia
9	Apr-07	Italy invades Albania
3	Sep-01	Germany invades Poland
9	Sep-03	UK & France declare war on Germany. Sea war begins. *Phoney War* on land & in air
	Dec-13	Battle of the River Plate
	April	10. Germany invades Denmark & Norway; 14. UK & France send troops to Norway
	May-10	Churchill becomes UK PM; Germany invades Holland, Belgium & France
	May 10- Jun-21	Battle of France
1	May-27-Jun-4	Dunkirk evacuation; UK cabinet decides to fight on
9	Jun-10	Italy declares war (North African war starts 4/7); Allies evacuate Norway
4	Jun-21	France capitulates
0	Jul-03	RN attacks French fleet in Algerian port
	Jul-Oct	Battle of Britain
	Sep-07	*Blitz* starts
	Oct-28	Italy invades Greece
	Nov-11	RN/FAA attack Italian navy at Taranto (blueprint for Japanese attack on Pearl Harbor)
	Apr-30	Greece evacuated
	May-27	*Bismarck* sunk
1	May-31	Crete falls
9		
4	Jun-22	Germany invades Russia
1		
	Dec-07	Japan attacks USA at Pearl Harbor, then invades Malaya and attacks Dutch possessions
	Dec-11	Germany declares war on USA
	May-4-8	Battle of Coral Sea (US/Australia-Japan)
1	Jun-4-7	Battle of Midway (US-Japan)
9	Jun-24	Tobruk falls. US send tanks to Egypt
4	June	US joins UK in aerial bombardment of Germany
2	Aug-19	British/Canadian raid on Dieppe
	Oct-23-Nov-11	Battle of el Alamein (UK-Germany/Italy)
	Nov-08	UK & US land in French colonies of Morocco & Algieria
		Long-range aircraft and escort carriers help close mid-Atlantic air gap
1	02-Feb	Battle of Stalingrad ends (began 23 August 1941)
9	13-May	Germany & Italy surrender in North Africa
4	24-May	U-boats leave Atlantic
3	09-Jul	UK & US invade Sicily
	03-Sep	UK & US land on Italian mainland
	Nov-Dec	Long-range fighters support US bombers attacking Germany
`1	Jan-27	Siege of Leningrad lifted (begun 8 September 1941)
9		
4	Jun-04	Rome liberated
4	Jun-06	D-Day (UK/USA/Canada invade Normandy)
	Aug-25	Paris liberated
	Jan-25	Battle of the Ardennes ends (begun Dec-16, 1944)
1	Mar-02	Berlin occupied by Russians
9	Mar-08	VE Day (European hostilities end)
4		
5	Aug-6 & 9	A-bomb dropped on Hiroshima & Nagasaki
	Aug-15	VJ Day (all hostilities end) WORLDWIDE PEACE

Battles

Note. In this table "UK" includes the forces of the British Commonwealth and her European allies.

BF Battle of France

B Br Battle of Britain

BLITZ

BATTLE of the ATLANTIC

N AFRICAN CAMPAIGNS

RUSSIAN CAMPAIGN

PACIFIC & SE ASIAN WAR

ITALIAN Campaign

W EURP

Foreword

Admiral the Rt Hon Lord West of Spithead GCB DSC PC

During the Second World War no nation on earth managed to integrate the best scientific and technological brains into their war effort to the extent of Great Britain. The results ensured victory and shortened the war.

In this labour of love Gwilym Roberts recognises the scientists, engineers and other experts whose ingenuity and invention, whose sheer brilliance, not only had vital application in war but contributed massively to future scientific development.

There have been numerous books about key inventions and their impact but not to my knowledge a chronological compendium of these breakthroughs laid against the campaigns in which they were used.

Roberts creates a broad definition of engineer to include scientists, designers, technical construction workers, maintenance personnel and users of technical equipment. He calls them 'technologists'. His biographical notes throughout the book give fascinating snapshots of the key players to whom our nation owes so much. These are men like Frederick Lindeman (Lord Cherwell) appointed by Churchill as special scientific advisor, Sir Henry Tizard, Sir Barnes Wallis, Sir Robert Watson-Watt; and many others, well and little known.

Few knew or know about the establishment of lateral thinking organisations such as the Military Intelligence Research Department (the Toyshop), the Admiralty Department of Miscellaneous Weapons development (the Wheezers and Dodgers) and S-Branch. Nor do many know that eighty years ago the Chain Home radar system was established along the south coast without which we would not have won the Battle of Britain. I must note, however, that the author falls into the trap of thinking that victory in the Battle of Britain stopped a German invasion. Captured German documents show conclusively that the German High Command had given up any idea of invasion owing to the strength of the Royal Navy and losses sustained by the German Navy in the Norwegian campaign.

The author highlights the many stunning achievements that assisted the course of the war in our favour: the countering of the magnetic mine, the development of the cavity magnetron, centimetric radar and the degradation of the German Knickebein system that was used for accurate bombing of British targets. Not least of these was the work of Bletchley Park.

One of the vignettes I found fascinating was from 1941 when a bomb destroyed Bank underground station creating a massive crater on a junction of six important roads. Within 90 minutes sappers and pioneers were clearing the site for the construction of a box girder bridge capable of carrying London buses nose to tail as illustrated in rather a good photograph. It was all completed in just four and a half days, something we can only dream about today.

The magisterial span of this book includes the vital contribution of women at war and the crucial improvements in medical science. Nor can we forget the iron men in small corvettes battling the North Atlantic gales and beating the U-boats of the cruel sea. Churchill famously stated "The only thing that ever really frightened me during the war was the U-boat peril". It was one of the longest battles in history with merchantmen sunk from the first to the last day of the war. Everything on land, at sea or in the air depended ultimately on its outcome. Roberts reminds us that the Battle of the Atlantic was highly technical and despite the bravery of those seafarers, it was won through the most innovative ideas and extraordinary schemes and inventions.

The book reveals the intricacies of D-day planning and logistics, from the Mulberry harbours, PLUTO and the less well-known DUMBO to Major General Percy Hobart's 'funnies' and advances in meteorological forecasting. It ends in the Pacific and the apotheosis of technical change, the Atom bomb.

Through this tour de force, which will enthral both layman and expert, Gwilym Roberts has achieved his aim of ensuring recognition for the myriad 'technologists' who made victory in World War II possible.

West of Spithead
August 2018

Preface

The initiative that prompted me to write this book occurred when I was sitting on a committee at the Institution of Civil Engineers that was tasked with proposing the names of a small number of eminent engineers whose names might be added to those already inscribed on the walls of the Institution's headquarters building in Westminster. Among the names considered was former Institution member R. J. Mitchell, the designer of the Spitfire fighter that played such a vital role in ensuring victory in the Battle of Britain. Although he was not chosen, I went on to appreciate that were it not for him and a few other scientists and engineers who made similar contributions before and during the war, we would not have enjoyed the liberties and lifestyles that we have had the good fortune to share for more than 60 years.

With Britain, America, Russia, and China all suffering defeats in the first years of the Second World War, the foundations for their later victories were laid by Britain, firstly, when she fought alone and won the Battle of Britain in 1940 and, secondly, by not losing the Battle of the Atlantic during the first two years of the war.

Although Britain and the Commonwealth were joined by Russia and the United States in 1941, the outcome of the war remained in the balance until 1942–43 when the Allies achieved success in a number of key battles, each of which turned the tide in their respective theatres, thereby leading to the final victory. In each of these key battles new technology played a decisive role.

Remarkable features of the war, especially when viewed from today's perspective, are the magnitude of many of the projects, the general tempo of the war, and the speed at which decisions were made, research and investigations undertaken, and machines and structures constructed.

Much has already been written by various people more knowledgeable than I about the inventions that were crucial to success in the vital battles of the Second World War. This book brings together and summarises such accounts and, set against the various battles and campaigns, gives an overview of the principal technical developments that influenced the course of the war. There are also biographical notes about some of the scientists and engineers whose achievements are described.

Scientists and engineers of all the combatant nations displayed remarkable ingenuity and invented and manufactured some exceptional machines and weapons. While the book mostly describes British achievements and individuals, some of those of her Allies are also described. In addition there is brief mention of some the significant German inventions that were made in extremely difficult circumstances.

While scientists undertake basic research and conceive new concepts it is engineers who convert such innovations into practical applications. Engineers and craftsmen are also the persons who fought in such large numbers in the technical branches of the armed services. For these reasons, I have used the term 'engineer' to embrace all the scientists, engineers, craftsmen, and other technologists mentioned in this book.

The book also gives me the opportunity to pay tribute to some of the major influences on my life and career, namely my school and university, the Royal Navy, and the wider engineering profession. I owe a great deal to each of them for having made it possible for me to follow the vastly interesting and varied career that it has been my privilege to have enjoyed.

Winston Churchill famously paid tribute to the fighter pilots who won the Battle of Britain when he said: 'Never was so much owed by so many to so few'. The small number of scientists, engineers and other technologists who made war-winning inventions and developments also deserve recognition; they are the real heroes of this book.

A very much shortened version of the book was presented to the Institution of Civil Engineers in July 2015 as their Smeaton Lecture for that year and can be viewed at ice.org.uk/Smeaton 2015.

Gwilym Roberts
Newick, East Sussex
2018

Chapter 1

GOD, CHURCHILL AND THE ENGINEERS

A winning combination

His Royal Highness the Duke of Edinburgh expressed it pithily: 'everything not invented by God was invented by an engineer. and 'They performed such an essential function,' he said to a Radio 4 audience in January 2016, 'that it was hard to imagine life without them,' and they held 'the key to the future of humanity and its ability to continue to thrive on the planet.'

As it is with the challenges of today, so it was in the punishing cauldron of the Second World War. We enjoy our present liberties thanks in large part to those scientists and engineers who helped bring about victory by inventing, designing, developing, and producing vital new systems, machines, structures, and weapons.

Many more engineers were in the armed forces, fighting valiantly alongside their non-technical colleagues and using the new equipment to maximum effect.

Britain's leaders, according to Sir Max Hastings, 'harnessed civilian brains and scientific genius to dazzling effect' during that conflict. 'Churchill's nation far surpassed Germany in the application of science and technology. Mobilisation of the best civilian brains, and their integration into the war effort at the highest levels, was an outstanding British success story.'

Definition of an Engineer

Scientists undertake basic research while **engineers** develop the results of such research into practical applications, which when built are generally maintained – and often operated – by **craftsmen**. Together these are often referred to as **technologists**.

In this book **Engineers** are the scientists, engineers and other technologists in both the fighting and civilian services who:

- Undertook basic research
- Conceived, designed, or developed new systems, machines, and weapons
- Manufactured or maintained such equipment
- Served in the technical units of the armed forces

More information about the various groups of military and civilian personnel involved and some associated details are contained in Appendix 1.

Lord Bowden went even further by claiming that 'All the courage of Fighter Command and the skill of the Few would have been wasted, had it not been for the timely installation of the primitive radar system which saved us in 1940.'

David Edgerton stated the following: 'The build-up of this empire of machines made Britain an exceptionally mobilized society where millions of people were making and using modern armaments on a huge scale. This warfare state was run by a wartime British government full of experts, of scientists and economists and businessmen.'

Winston Churchill, while First Lord of the Admiralty in February 1940, offered this encomium in Parliament: 'I wish this afternoon to pay my tribute to the Engineering Branch [of the Royal Navy] ... the man around the engine without whom nothing could be done, who does not see the excitements of the action and does not ask how things are going, but who runs a very big chance of going down with the ship should disaster come.'

Britain established the most effective interaction between technologists and its military of any country that fought the war. The seeds of this success were sown during the rearmament programme of the pre-war years, when the Royal Society and other professional bodies identified key technical activities which in the event of conflict should become reserved occupations, freeing those engaged in them from serving in the armed forces.

They also prepared a register of scientists and engineers whose work could be of particular benefit which by 1939 contained some 5,000 names. In addition, state bursaries were awarded to over 50,000 young people (including the author) to enable them to study a technological subject before entering the fighting services or going into industry.

The Impact of Churchill

The technical input into the war effort was transformed when Churchill became Prime Minister in May 1940. He not only actively supported technology and innovation but 'brought to the pursuit of science the same boundless energy with which he prosecuted every other aspect of the war.'

He appointed his friend and confidante, Frederick Lindemann (later Lord Cherwell) FRS, a former head of Oxford's Clarendon Laboratory, as his special scientific advisor. *Scientific American* has described him as 'the most powerful scientist ever'. The two met almost daily, and Lindemann attended meetings of the War Cabinet.

Murderous Thoughts

Although Lindemann's activities were almost entirely beneficial to the war effort, he pursued a number of impracticable ideas and his appointment was intensely disliked by a number of the other leading scientists who believed he had a Rasputin-like influence on Churchill.

Lord (Solly) Zuckerman, the naturalist turned operational research expert, remarked that Lindeman was 'the only person ... whom I have ardently wished to murder'!

Cherwell, Lord PC CH FRS (1886–1957)

Although born in Germany, Frederick Lindemann was a British national because his father, a German émigré, had acquired British nationality; his mother was American. Educated in Scotland and Germany, he inherited wealth and was a teetotal non-smoker, an accomplished pianist, and an international-standard tennis player. Working in Berlin as a physicist just before the First World War, he managed to return to Britain and having learned to fly with the RFC, he worked out how an aeroplane could be taken out of a spin.

After the First World War he became head of Oxford's Clarendon Laboratory, which he transformed from a 'museum piece' into an acclaimed research body. He became a close friend of Winston Churchill in 1921 and 11 years later the two went on a road trip to Germany, where they were dismayed with what they saw. When Churchill was out of office Lindemann advised him on scientific matters; Churchill arranged for Lindemann to become a member of the Committee for the Scientific Study of Air Defence (whose chairman was Henry Tizard). However, Lindemann's contributions were disruptive and the Committee disbanded and then reformed without him.

When Churchill became Prime Minister, Lindemann was appointed as the Government's senior scientific advisor. Known as 'The Prof', he attended meetings of the War Cabinet, met Churchill on a daily basis, and accompanied him on his overseas journeys. In addition to being closely associated with the specialist department MD1, Lindemann established a distinct statistical branch known as 'S-Branch'.

A number of other leading scientists were highly critical of his appointment and of the influence he had on Churchill. In 1951–52 he served in Churchill's post-war administration. He was appointed a Companion of Honour in 1953 and elevated to the peerage in 1972 when he assumed the title of Lord Cherwell. He died peacefully a year later.

Churchill and Lidemann witness the testing of a new weapon, *IWM*

'Churchill was a great enthusiast for science and machines,' David Edgerton wrote, 'particularly in relation to war, in a country where the elite, and especially the old aristocratic elite from which Churchill came, were thought to be either above such matters or sunk in rural idiocy.'

He went on to quote Oliver Lyttleton, the Minister of Production in the Cabinet: 'One of Churchill's most important qualities as war leader was his eager readiness to listen to new, sometimes fantastic, ideas thrown up by scientists, engineers and academic figures.'

Churchill was a competent inventor himself. Commander Sir Charles Goodeve FRS RNVR, the Admiralty's senior scientist, said he was 'an inventor of no mean repute'. When First Lord of the Admiralty during the First World War, Churchill was the 'key figure behind the invention of the tank', which was originally called a 'landship'.

When holding that appointment again at the start of the Second World War, he developed and promoted Nellie, a giant trench-digging excavator which would enable troops to advance on enemy positions while protected so as to provide a means of 'breaking a deadlock on the French front without repetition of the slaughter of the previous war'. He also promoted the development of floating mines for dropping into German rivers.

The conception, research, and development of new weapons and machines were undertaken by the research establishments of the three military ministries; by academia and industry; and by two small specialist departments enthusiastically supported by Churchill. These were the Ministry of Defence 1 (MD1), known colloquially as 'Winston Churchill's Toyshop', and the Admiralty Department of Miscellaneous Weapons Development (DMWD), otherwise known as 'the Wheezers and Dodgers'. These two departments developed some 50 significant inventions including limpet mines, the Navy's Hedgehog depth charge launcher, and the infantry's PIAT anti-tank mortar.

Inside the Toyshop

Shortly before the outbreak of war the War Office had established a new department under the direction of Lieutenant-Colonel (later Major-General) Joe Holland RE. This was called Military Intelligence (Research) (MIR) and was intended to work closely with the Foreign Office on military intelligence matters. Assisting Holland were Major (later Major-General Sir) Millis Jefferis RE and Stuart (later Colonel Stuart) Macrae, the then editor of the magazine *Armchair Science*.

The organisation became known to Churchill, then the First Lord of the Admiralty, when he came up with his idea for creating floating mines – a task which its boffins successfully achieved and demonstrated to various British and French VIPs. After he became Prime Minister he continued to take a close interest in the department's activities, and he overrode senior officials who wanted it incorporated into the Ministry of Supply. He ruled in November 1940 that it should become the first subsidiary department of the Ministry of Defence – hence MD1. It was then effectively under the direct control of Churchill and Lindemann, who made weekly visits to the department.

The department's unofficial history, written by Macrae's son and entitled *Winston Churchill's Toyshop*, states that it designed 26 entirely new weapons which went into quantity production and which ranged from small booby-traps to heavy artillery, aircraft bombs, and naval mines.

Initially housed in a small office in the War Office in Whitehall, the department moved in 1940 to Portland Place, but after that building was damaged by bombing a few months later additional premises were found at The Firs, Whitchurch, near Aylesbury in Buckinghamshire. This was a country house large enough to provide offices and accommodation, and had stabling that could be converted into workshops. A telephone network was installed at Portland Place which enabled direct communication from both stations with the War Cabinet and various War Office departments.

In his history of the war, Churchill wrote: 'This was … no time to proceed by ordinary channels in devising expedients. In order to secure quick action, free from departmental processes, upon any bright idea or gadget, I decided to keep under my own hand as Minister of Defence the experimental establishment formed by Major Jefferis at Whitchurch.

'While engaged upon the fluvial mines in 1939 I had had useful contacts with this brilliant officer, whose ingenious inventive mind proved, as will be seen, fruitful during the whole war. Lindemann was in close touch with him and me. I used their brains and my power.'

A demonstration range, explosive filling sheds, pools for underwater experiments, and production units for certain weapons were built in the grounds and women were

Technological Support

Churchill, in addition to being Prime Minister, assumed the title of Minister of Defence – even though there was no Ministry of Defence as such. In addition to leading the Cabinet, he presided over the War Cabinet which comprised himself and four (later six) senior ministers. Among the bodies that reported to the Cabinet was its Scientific Advisory Committee.

bussed in from a hostel to work the machines. A range at Risborough was also used to demonstrate weapons to the Prime Minister and other VIPs.

As new weapons were produced, Macrae, using his journalistic experience and contacts, oversaw the writing of instruction manuals which included 'exploded' diagrams of the weapons. Found to be more understandable than the line diagrams used in conventional War Office manuals, such diagrams came eventually to be used in all manuals.

Although MD1 originated as a War Office department, its fame and reputation were such that both the Royal Navy (RN) and the Royal Air Force (RAF) consulted it on aspects of weapon design amongst other things. As with DMWD, many ideas from would-be inventors were passed to MD1 for appraisal and testing, activities that, in the early days when they were extremely short-staffed, diverted them from their main task.

See Appendix 3 for more details.

The 'Wheezers and Dodgers'

A remarkable department was established in the summer of 1940 and led by a senior RN officer who reported directly to the Board of Admiralty. Originally established to design better anti-aircraft protection for RN and Merchant Navy ships, its first title was the Admiralty Anti-Aircraft Weapons and Devices Department, but its name was changed to the Department of Miscellaneous Weapons Development (DMWD) after it became involved with the development of devices to attack U-boats. Not having comparable organisations of their own, both the Army and RAF referred problems and ideas to DMWD.

Many recruits to the department were transferred from HMS *King Alfred*, the training establishment for potential RNVR officers based in Hove, Sussex. Virtually all DMWD staff were appointed RNVR Special Branch officers; as such, they were mostly ignorant of Admiralty procedures for procurement and disbursement and found it easier to circumvent red tape than RN officers would have done. Fortunately, both the Admiralty's Directorates of Scientific Research and of Naval Accounts adopted a tolerant attitude to DMWD's unorthodox activities.

Gerald Pawle wrote: 'It was the complete freedom to experiment, the freedom to tackle unorthodox projects in an unorthodox way, which was the basis of DMWD's success. And it was greatly to the credit of the Admiralty that they allowed such a free hand to an organisation whose approach to most problems must have seemed revolutionary in the extreme.'

These sentiments were echoed by the Admiral of the Fleet Lord Fraser of North Cape who said, 'Their job could only have been done if they were unhampered by routine work.'

For the first two years of its existence its senior technical officer was Commander Goodeve FRS RNVR, whose principal deputies were Commander Richardson RNVR, a former scientific colleague of Goodeve's at Imperial College, and Lieutenant-

A Penny for your Thoughts

Following the precedent set after the First World War, at the end of the Second World War a Royal Commission on Awards to Inventors was established which considered applications from people who had made inventions that had been used during the war; 359 applications were considered.

Military personnel and civil servants who developed such equipment as part of their duties were excluded unless the inventions were 'of such exceptional brilliance and utility that some award might nevertheless be justified'.

Among the cases which were deemed to be so exceptional were those of Sir Robert Watson-Watt who was awarded £50,000 for his invention of radar (a number of his colleagues were also recognised); Sir Donald Bailey, who was awarded £12,000 for his invention of the Bailey Bridge; and Air Commodore Sir Frank Whittle, who received £100,000 for his invention of the jet propulsion engine. To many this latter award seems a trifle odd as, brilliant and ultimately beneficial though the invention was, it came about too late to play a significant role before the end of hostilities. It was however explained by the Commission in a footnote: 'Air Commodore Sir Frank Whittle was not in fact a claimant. The commission investigated the case at the special request of the Ministry of Supply in agreement with the Air Commodore and the Treasury.'

The fact that an applicant did not receive an award often meant that it did not comply with the guidelines, rather than it was not worthy of an award (a number of the engineers involved in designing the Mulberry harbours had their applications rejected).

Commander N. S. Norway RNVR – better known as the author Nevil Shute – an engineer who had worked on airship design pre-war.

In the first year of the war, when he was still attached to HMS *Vernon* and before DMWD had actually been established, Goodeve had conceived and developed the means of countering the magnetic mines that caused the loss of so many ships in the early months of the war (see Chapter 3).

The department was disbanded shortly after the end of the war, but a reunion dinner was held at Simpson's in the Strand in May 1953. It was chaired by Goodeve, who by then had been knighted for his wartime services. The value and extent of their contribution to the war effort is reflected in the fact that Goodeve and four other members all received monetary awards from the Royal Commission on Awards to Inventors.

Details of some of the Department's inventions are given in Appendix 4.

Some Other Strange Beasts

'Winston Churchill's Toyshop' and the 'Wheezers and Dodgers' weren't the only unorthodox departments to marshal the country's brightest brains.

S-Branch was a small statistical organisation established at Marlow by Lindemann which reported directly to Churchill. It scrutinised the performance of the regular ministries and, having analysed data from a variety of sources, produced easily-understood reports and charts thereby enabling key aspects of the war's progress and the nation's resources to be readily evaluated.

The importance and significance of these presentations are highlighted by the charts now on display in the Cabinet War Rooms. Inter alia these show the tonnage of shipping lost each month compared with new construction and the weight of bombs dropped by Germany on the UK compared with that dropped on Germany.

Although S-Branch often caused tensions between government departments it enabled the War Cabinet to make quick decisions based on accurate data and was undoubtedly an important component of the organisation behind the successful prosecution of the war.

The Allied Central (Photographic) Interpretation Unit (ACIU) was based at Medmenham, Buckinghamshire. Originally the RAF Photographic Interpretation Unit, it became the Anglo-American Allied Central Interpretation Unit after the entry of the Americans into the war.

During 1942 and 1943 the unit gradually expanded and was involved not only in the planning stages of practically every operation of the war but in every aspect of intelligence. In 1945, the daily intake of material averaged 25,000 negatives and 60,000 prints. By VE Day it employed 1,700 personnel, a large number of whom were women, and the print library, which documented and stored worldwide cover, held 5,000,000 prints from which 40,000 reports had been produced.

Combined Operations Headquarters was a department of the War Office established in July 1940 to harass the Germans on the European continent by means of amphibious raids carried out by commandos. Its first director was Admiral of the Fleet Sir Roger Keyes, who had won distinction in the First World War for leading a raid on the Dutch port of Zeebrugge. (He was succeeded in October 1941 by Admiral Lord Louis Mountbatten followed by Major General Robert Laycock in October 1943.)

Bizarrely, the organisation became the godfather to a scheme named Project Habakkuk, which envisaged the manufacture of reinforced icebergs for use as floating aircraft carriers. It was conceived by the eccentric inventor Geoffrey Pyke, an advisor to Mountbatten, and it was named after the minor Hebrew prophet whose eponymous book in the Bible includes a phrase reflecting the project's ambitious goal: 'Be utterly amazed, for I am going to do something in your days that you would not believe, even if you were told.' (Hab. 1:5 New International Version).

A news correspondent in Germany at the start of the First World War, Pyke had been interned but had managed to escape – a feat which earned him a measure of notoriety. Between the wars he had invested heavily in the stock market, founded a school, and become bankrupt. Following the German invasion of Norway he conceived a novel method for transporting troops across snow fields using screw-propelled vehicles called Ploughs. The idea was taken to Mountbatten who thought it worthy of further investigation. He took Pyke on to his staff, both for his original ideas and because he prompted other staff members to think less conservatively. The concept was then actively pursued by the Americans and Canadians but was eventually superseded by the Canadian Weasel and the American M29 tracked personnel carriers.

In mid-1942 Pyke was asked to investigate problems concerned with the icing of ships in Arctic waters. This prompted his lateral-thinking brain to consider related matters, in particular whether artificial icebergs could be used as aircraft carriers that could help close the mid-Atlantic air gap or support an amphibious landing on a coastline, such as the Bordeaux region of France – then beyond the reach of land-based aircraft.

To do this he conceived the manufacture of a new material made from wood, pulp, and frozen water which became known as Pykrete and was extremely hard and slow-melting. Examples were given to Mountbatten, and by him to Churchill. Both of them actively supported the idea, which was considered by the combined chiefs of staff at their meeting in Ottawa in August 1943.

An Anglo-American-Canadian committee was established to investigate further, though by the end of the year the need for such a project had been nullified both by Allied successes in the Battle of the Atlantic and by the decision to invade France through Normandy, which was within reach of UK land-based aircraft.

Sunday Soviets

Britain organised collaboration between the military, scientists, and industry supremely well – discussions between the respective participants were open and frank with the objective of finding not the perfect solution, but one which worked and could be applied effectively and quickly in as many applications as possible.

This willingness to discuss ideas and problems was well illustrated by the 'Sunday Soviets' who met on Sunday mornings at the Telecommunications Research Establishment (TRE) in Malvern. Originating with discussions between senior RAF officers and Robert Watson-Watt and Jimmy Rowe (the scientists who developed radar), they developed into regular meetings of an informal nature and were attended by senior officers from all three services. Here the operations of a whole command could be discussed, and junior personnel who had a useful contribution to make, whether technical or military, were invited to attend and encouraged to speak.

It has not been possible to ascertain why they were called Soviets, a word often associated with the Russian Communist regime.

Tots and Quots

Professor Solly Zuckerman had established in 1931 a London-based dining club called the Tots and Quots, which included many leading members of the British scientific establishment. Its name was derived from the Latin tag *Quot homines, tot sententiae* ('As many opinions as there are men').

Zuckerman,Lord Solly OM KBE FRS (1904–93)

A zoologist and Operational Research pioneer, he was born to Jewish immigrants in South Africa. After studying medicine at the University of Cape Town and later attending Yale University, he went to London in 1926 to complete his studies at University College Hospital Medical School. He began his career at the London Zoological Society in 1928 and worked as a research anatomist until 1932. He taught at the University of Oxford in 1934–45, in which time he was elected FRS. In 1931 he established the Tots and Quots scientific dining society. During the Second World War he undertook various research projects, advised the RAF, and proposed that in the build-up to the Normandy landings the Allied air forces should target the French railway network, in particular its locomotives and locomotive repair facilities.

In June 1940 a meeting was attended by Allen Lane, the founder of Penguin Books, who was so fascinated by the discussion about what scientists could do to help win the war that he offered to publish the arguments. Eleven days later 25 scientists delivered their manuscripts, and in late July 1940 a 140-page Penguin Special entitled *Science in War* was published, opening with the stark message, 'The full use of our scientific resources is essential if we are to win the war. Today they are being half used.'

Our American Allies

From the beginning of the war Henry Tizard, the rector of Imperial College and chairman of the Committee for the Scientific Study of Air Defence, had been considering how to cultivate ties with American scientists with the objective of 'bringing American scientists into the war before their government'. As a result, in February, 1940 Professor Hill went to the USA to discuss with American scientists possible areas of collaboration. Encouraged by the discussions, he and Tizard set about obtaining Cabinet approval for the dispatch of a formal scientific mission to the USA to share the UK's military and scientific secrets with the Americans.

The result was that in September 1940, while the Battle of Britain was at its height and well over a year before the Americans entered the war, Tizard himself led a mission to the USA to disclose Britain's most carefully guarded secrets. These included designs and hardware for anti-submarine detection, explosives, gyro-gunsights, jet propulsion, micro-pump valves, proximity fuses, radar, cavity magnetrons, rockets, and a possible atomic bomb. Of these it was the cavity magnetron which, by increasing by a factor of

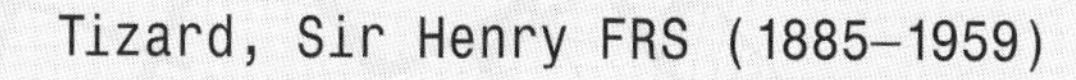

Tizard, Sir Henry FRS (1885–1959)

He was educated at Westminster and Oxford and, as a physical chemist, was elected a Fellow of Oriel College, Oxford, in 1911. In the First World War he joined the RFC, where he undertook aerodynamic observations, worked on bombsights, and, having learned to fly, became his own test pilot.

Upon returning to Oxford in 1919 he was appointed a reader in chemical thermodynamics; working on the performance of petrol engines and, he developed octane numbers for rating fuel. In 1919 he was made a member of the Aeronautical Research Committee and the following year he became assistant secretary of the Department of Scientific and Industrial Research which was responsible for co-ordinating the scientific work of the defence and civil departments. In 1933 he was appointed chairman of the Aeronautical Research Committee and two years later chairman of the new Committee for the Scientific Study of Air Defence. These were the bodies that first discussed Robert Watson-Watt's idea for detecting the presence of aircraft by radio beams. The following year he and others resigned from the latter committee as a result of Lindemann's manoeuvres but he continued to take an active part in the development of radar. He encouraged work at the University of Birmingham that led to the invention by John Randall and Harry Boot of the cavity magnetron. This provided the basis for, among other things, centimetric radar, thereby making airborne radar interception possible. In September 1940, with the Battle of Britain at its height, he led a mission to share with the Americans Britain's scientific and military secrets which proved to be one of the key events in forging the Anglo-American technical alliance in the Second World War. He later played a key part in winning support for Frank Whittle's jet engine. In 1929–42 he was rector of Imperial College London, prior to becoming president of Magdalen College, Oxford, for the following five years.

Leaving Oxford in 1947, he returned to Whitehall as chairman of the Defence Research Policy Committee and a member of the Advisory Council on Scientific Policy. He retired from government service in 1952 when he became pro-chancellor of the University of Southampton. He died in Hampshire in 1959.

a thousand the transmitting capabilities of American radar, convinced the Americans that the British were disclosing everything they knew.

Tizard's package was described by an American as 'the most valuable cargo ever to reach our shores', while to a Briton 'the decision to (disclose) all the UK's secrets, showed great wisdom and boldness'. On their safe arrival in Canada, the two scientists

Hill, Archibald V. CH OBE FRS (1886–1977)

A. V. Hill, as he was generally known, was a physiologist and one of the founders of biophysics and of operational research. A Cambridge graduate, he served in the Army in the First World War undertaking ballistics research; his peacetime work was on muscles, for which he was awarded a Nobel Prize in 1922. He was a professor at University College London from 1923 to 1951. In 1935 he worked with Patrick Blackett and Henry Tizard on the committee that gave birth to radar. In 1933, he was a founder member and Vice-President of the Academic Assistance Council (later the Society for the Protection of Science and Learning) which, in particular, assisted refugee Jewish scientists to re-establish themselves in the British academic establishment.

With Blackett he resigned from Tizard's Aerial Defence Committee in 1935 because of Lindemann's manoeuvrings and later was highly critical of his influence on Churchill. Prior to the 1940 Tizard Mission to the USA, Hill went to Washington to sound out his American contacts on their likely reaction to such an initiative; his positive report was a factor in persuading the British Government to proceed with the mission. He knew many leading scientists well, had many influential contacts and, *inter alia*, recommended Blackett's appointment to Anti-Aircraft Command. He served as an independent MP for Cambridge University in 1940–45. He took part in many scientific missions to the US. He was appointed OBE in 1918 and also elected FRS; he became a CH in 1946.

who conveyed the hardware across the Atlantic, John Cockcroft and Taffy Bowen, were shown a pistol by the ship's captain who told them that he had been instructed, had mishap befallen the ship, to ensure by any means at his disposal that they were not to be taken alive by the enemy.

American military science was the responsibility of the National Defense Research Committee, whose establishment was authorised by President Roosevelt after reading a paper by Vannevar Bush, the head of the Carnegie Institution. Bionote in Chapter 10.

As the USA and UK had a similar approach regarding the scientist-military relationship there was close Anglo-American technical collaboration throughout the war, with nationals from both countries working alongside each other. Often devices developed in the UK were wholly manufactured in the USA; for many weapons and machines this became Britain's arsenal, with American productive technology manufacturing most of the ships, tanks, aircraft, and armaments that eventually overwhelmed the Axis powers (see Chapter 6).

The fact that Anglo-American technologists had virtually a free hand to develop novel equipment, coupled with the immense productive capacity of the USA, meant that Anglo-American leaders were blessed with equipment vastly superior to that of their enemies.

After a mere 21 years of uneasy peace, the world was at war again – a war in which technology would play a crucial role in securing victory for Britain and her allies. Meanwhile in a mansion in Buckinghamshire, many brains were already hard at work trying to decipher the enemy's messages...

BLETCHLEY PARK TIMELINE

The Enigma Encrypting Machine

1920s German engineer Arthur Scherbius invents the Enigma electro-mechanical encrypting machine. Used commercially and also by several governments, including Germany, to protect military and diplomatic communications.

1932 Polish mathematician and cryptologist Marian Rejewski partly breaks the German encryption system. French spy working in Germany acquires further information which the French passed to the Poles who are then able to read German messages and build their own replica machines.

1938 Germans develop a more complex machine which the Poles are unable to read.

1939 In July the Poles inform French and British cryptologists of their work and, after the outbreak of war, they and their equipment are evacuated to France. They also give one of their replica machines to British cryptologists.

1940 Following the fall of France, the Poles move first to Vichy, France, and then to Britain in November 1942, where they collaborate with British cryptologists.

British Decrypting Work

1938 Government Code and Cypher School (GC&CS) acquires Bletchley Park, Buckinghamshire.

1939 Recruitment of academics, mathematicians etc. by GC&CS. Separate huts established to tackle the different systems used by each of the German armed forces.

1940 HMS *Griffin* captures German trawler. Some naval codes temporarily broken.

1941 The Germans start using the more sophisticated Lorenz cypher machines. The British name deciphered Lorenz messages Tunny.

Alan Turing and Gordon Welchman develop a Bombe to speed up decoding procedures.

In May, a RN ship obtains Enigma machine and code books from a sinking U-boat.

In December Germany declares war on USA and American cryptologists start working with their British counterparts at Bletchley Park.

1942–43 Post Office engineer Thomas Flowers builds Colossus which implements Turing/ Welchman proposals and enables Lorenz signals to be read more speedily.

1945 Government orders destruction of all machines.

Chapter 2

A BATTLE OF WITS

Bletchley and beyond

In 1938 the British Government Code and Cypher School (GC&CS) acquired Bletchley Park, Buckinghamshire (in present-day Milton Keynes) as its headquarters. On the outbreak of war it recruited a number of top-level academics – particularly mathematicians and linguists, but also historians, chess champions, and crossword addicts – to work on decrypting German radio intelligence signals picked up by intercept stations (see Y-service). The German signals were encoded using the highly sophisticated Enigma and Lorenz encrypting machines, amongst others, which produced coded messages that the Germans regarded as unbreakable.

Thanks to incredibly brilliant decrypting work by the Bletchley Park cryptologists, coupled with the development of electro-mechanical computing machines, the Enigma (and later, even more sophisticated systems) was broken. As a result, the Allied High Command was informed of German plans and dispositions, often within a few hours of their radio transmission.

Many women worked at Bletchley Park, some as cryptologists but several more providing various support services. In particular there were many hundreds of Wrens (WRNS – Women's Royal Naval Service) who assisted in the operation of the decrypting machines. A few Wren officers also served as decrypting officers in the troopships converted from the fast peacetime passenger liners whose speed enabled them to cross the oceans unescorted. If Bletchley Park decrypts suggested U-boats might be approaching, signals were sent instructing them to change course.

Down Memory Lane

At the end of the war the decoding machines and many relevant papers were destroyed. (The machines now on display at the museum in Bletchley Park are recently constructed replicas.) Due to the loyalty of those who worked there during the war, all of whom were bound by the Official Secrets Act, the whole Bletchley Park operation remained unknown to the public until the 1970s and 1980s when there started to be leaks about its activities; following this some of those involved wrote their memoirs. Lately its operations have been described in a number of books and portrayed in films and television programmes. Bletchley Park's present-day successor is the Government Communications Headquarters (GCHQ) at Cheltenham.

The decoded information, called Ultra, had a profound effect on the outcome of the Battles of the Atlantic and Alamein, and the North Africa and north-west Europe campaigns. It has been claimed that the knowledge gained through Ultra shortened the war by two years. After the entry of the USA into the war American cryptanalysts worked alongside their British counterparts at Bletchley Park. 'It was thanks to Ultra,' Churchill is credited with saying, 'that we won the war.'

Enigma

Courtesy of Greg Goebel

Originally invented by a German engineer around 1920, early Enigma encoding machines were used commercially but were later adopted by the German military who also upgraded them. The Poles were also working on them; by December 1932 they had broken the German cyphers. They also built replica machines and, five weeks before the outbreak of the Second World War, they gave replica equipment to the British GC&CS and their French equivalent together with details of their decryption techniques.

The Enigma machines worked on electro-mechanical principles and had five rotors, each with a variety of settings which were generally changed on a daily basis, meaning time was always of the essence – it was calculated that there could have been 159×10^{18} possible daily keys.

Separate coding systems were used by each of the German services, adding to the difficulties of their decryption. As the numbers working at Bletchley Park increased, wooden huts were built in the grounds, with each hut's occupants dealing with separate aspects of the decrypting operation. Hut 6 under Gordon Welchman was responsible for dealing with Army and Air Force codes, while Navy codes were the responsibility of Hut 8, which was initially controlled by Alan Turing and subsequently by Hugh Alexander.

In the early days the successful cracking of a code relied to a certain extent on insights and inspired guesses by the cryptologists as to the likely behaviour of the German operatives of the machines. Some frequently used insights were named after the person who first proposed them e.g. the Herivel tip, or Herivelismus, after John Herivel and Parkerismus after Reg Parker.

Outstanding work by the GC&CS cryptanalysts enabled some messages to be broken as early as 1940. Further clues as to the workings of the Enigma machines and of the naval codes used were obtained by the capture of German weather boats off Iceland in 1940–41; however, as code settings were changed each month they could be broken only for a limited period. Then, in May 1941, U-boat *U-110* attacked a convoy though depth charges fired by the escorting vessels forced her to surface, whereupon her crew abandoned ship. Instead of ramming and sinking her, the captain of HMS *Bulldog* ordered one of his officers, Sub-Lieutenant David Balme RN, to be rowed over, board the submarine, and retrieve anything of interest. Among the prize items of equipment he took back to *Bulldog* were an Enigma machine and current code books, which were then rushed to Bletchley Park in great secrecy. (Balme was awarded the DSC.) Next, in October 1942, *U-559* was sunk in the eastern Mediterranean. Before she went under, three crew members of the attacking ship, HMS *Petard*, were able to board her and recover the code books. Sadly, the submarine sank suddenly, taking two of the boarding party with her.

Lorenz and Tunny

In mid-1941 the Germans started using the more sophisticated Lorenz cypher machines; some 12 months later, the German High Command began to use the system for its high-level communications with its Army Commands. The Lorenz signal traffic was known as Tunny by the Bletchley Park cryptanalysts. As the result of an error by a German transmitter in late August 1941, by January 1942 John Tiltman and Bill Tutte had worked out the complete logical structure of the cypher machine, thereby enabling Tunny to be decoded. This remarkable piece of reverse engineering was later described as 'one of the greatest intellectual feats of World War 2'.

Tiltman, John Hessell Brigadier CMG CBE MC (1894–1982)

A British Army officer, he moved to intelligence work after being wounded during the First World War, initially working with the Indian Army and then at the Government Code and Cypher School (GC&CS). His intelligence work was largely connected with cryptography, and he showed exceptional skill at cryptanalysis. He was considered one of Bletchley Park's finest cryptanalysts on non-machine systems and worked with Bill Tutte on the cryptanalysis of the Lorenz cypher. In 1944, he was appointed deputy director of GC&CS and continued there until 1949 when he moved to the US Army Security Agency as a liaison officer. He eventually became a consultant and researcher at the US National Security Agency. In September 2004, he was inducted into the 'NSA Hall of Honor', the first non-US citizen to be recognised in that way. The NSA commented:'His efforts at training and his attention to all the many facets that make up cryptology inspired the best in all who encountered him.'

Tutte, William (Bill) Thomas OC FRS FRSC (1917–2002)

The son of a gardener, he became a British, and later a Canadian, mathematicianand code-breaker. Having graduated with a degree in chemistry from Trinity College, Cambridge, he worked at Bletchley Park, initially on Italian Navy codes but from mid-1941 on Tunny, the signal traffic generated by the Lorenz coding machine. His brilliant and fundamental deciphering work led to the code being broken and, eventually, to the creation of the Colossus computing machine. Post-war he emigrated to Canada where he held senior positions at the universities of Toronto and Waterloo. He had a number of significant mathematical accomplishments, including foundational work in the fields of graph and matroid theories.

Turing, Alan (1912–54)

A mathematician and computer scientist, Turing was appointed a fellow of King's College, Cambridge, in 1935 at the early age of 23. At the outbreak of war he was recruited to the Government's code-breaking staff at Bletchley Park where he was instrumental in breaking the German naval codes. With Gordon Welchman he designed the Bombe decoding machine and then with Tommy Flowers he designed and developed Colossus, the world's first computer. Post-war he was appointed to a chair at the University of Manchester, where he continued his computer developments. Turing was prosecuted for homosexual acts in 1952, when such behaviour was still a crime in the UK. He accepted treatment with oestrogen injections (chemical castration) as an alternative to prison. He committed suicide in 1954 and was granted a posthumous pardon in 2013.

As laborious manual methods necessarily had to be used, the decrypting process was extremely slow. Turing and Welchman therefore set about developing machines which could undertake the work speedily and accurately. The first was the Turing-Welchman Bombe, of which a number were manufactured in 1941; in case Bletchley Park should be bombed, many were installed at outstations.

Then came the electronic Colossus computing machine, in whose development and manufacture Thomas Flowers, a Post Office engineer, was very largely responsible. It used some 1,800 thermionic valves – a substantial increase on the previous most

Welchman, Gordon OBE (1906–85)

A mathematician, code-breaker, and computer scientist; although a graduate of Trinity College, Cambridge, he was appointed a fellow and dean of Sidney Sussex College in 1929. On the outbreak of war he joined the Government's code-breaking school at Bletchley Park and recruited a number of fellow Cambridge mathematicians to work with him. He first headed the Hut 6 team which was charged with deciphering German Army and Air Force codes. With Turing he went on to develop the Turing-Welchman Bombe that enabled German codes to be broken and from which modern computers developed.

He emigrated to the USA in 1948, and at MIT taught the first computer course in the United States. He then worked for Remington Rand, Ferranti, and the MITRE Corporation on tactical communications systems for the US military. He became a naturalised American citizen in 1962 and retired in 1971, but was retained as a consultant. His book, *The Hut Six Story*, was published in 1982. The National Security Agency disapproved of this and he lost his security clearance and, in consequence, his consultancy with MITRE. He was also forbidden from discussing either the book or his wartime work with the media. Welchman died in 1985 and his final conclusions and corrections to the story of wartime code-breaking were published posthumously in 1986 in a paper entitled 'From Polish Bomba to British Bombe: the birth of Ultra'.

Wrens operating a Colossus machine. PD National Archives FO850/234

Flowers, Thomas MBE (1905–98)

The son of a bricklayer, Flowers served an apprenticeship at the Royal Arsenal, Woolwich. While there he obtained an electrical engineering degree from the University of London. He then joined the Post Office as an electrical engineer at their Dollis Hill research station, where he explored the use of electronics for telephone exchanges. In 1942 he was appointed to work with Alan Turing to build a decoder for the relay-based Bombe machine that Turing had developed to help decrypt Enigma codes. Although the decoder project was abandoned, Turing was impressed with Flowers' work, and in February 1943 they began working on developing an automated decoding system for Lorenz cyphers. After initial work with a machine called Heath Robinson, Flowers proposed an electronic system (called Colossus), which was successful and had a major influence on the war's proceedings. Post-war he became the Post Office's chief engineer and also worked with the National Physical Laboratory in computer development. His vital work at Bletchley Park was not fully appreciated until many years after the end of the war.

complicated electronic device which had only used about 150. Some of the senior Bletchley Park management were unconvinced that Flowers' idea would work, saying he was 'squandering good valves', and his funding was cut off, forcing him to pay for the subsequent work himself. However, the project did work, and the first machine became operational in December 1943, thereby enabling Lorenz messages to be read. By the end of the war there were ten in service; they were the precursors of today's computers.

The organisational arrangements at Bletchley Park reflected the British approach in successfully harnessing brilliant minds to support the war effort. Turing, Welchman, and their colleagues had the ideas, but they needed sympathetic minders such as Commanders Dennison and Travis, both RN, to make their work effective and to provide the organisational framework. It is doubtful whether anything similar occurred anywhere in Germany during the war.

Y-Service

Operating out of some 30 separate stations, personnel from the Royal Signals and other agencies listened into German wireless traffic; later from stations in India and elsewhere in the Far East, they listened into Japanese signals. Many amateur ('ham') radio operators supported the work of the 'Y' stations, being enrolled as 'Voluntary

Interceptors'. Much of the traffic intercepted by these stations was recorded by hand and sent to Bletchley Park on paper by motorcycle couriers or, subsequently, by teleprinter over Post Office landlines. A large house called Arkley View on the outskirts of Barnet acted as a data collection centre at which traffic was collated and passed to Bletchley Park. It also acted as a 'Y' station.

In addition to wireless interception, specially constructed 'Y' stations also undertook High Frequency Direction Finding (HF/DF, known colloquially as 'Huff Duff') on enemy wireless transmissions. This became particularly important in the Battle of the Atlantic (1939–1945), where locating the positions of U-boats became a critical issue. Admiral Dönitz told his commanders that they could not be located if they limited their wireless transmissions to under 30 seconds, but skilled HF/DF operators were able to locate the origin of their signals in as little as six seconds.

The land-based DF stations preferred by the Allies operated on the Adcock antennae system, which consisted of a small central operators' hut surrounded by four 10-metre-tall aerial poles, usually placed at the four compass points. Aerial feeders ran underground and came up in the centre of the hut; these were connected to a direction-finding goniometer and a wireless receiver that allowed the bearing of the signal source to be measured. In the UK some operators were in underground metal tanks. These stations were usually located in remote places, often in the middle of fields. Traces of Second World War D/F stations can be seen as circles in the fields surrounding the village of Goonhavern in Cornwall.

Far Eastern Codes

The Japanese had different coding systems for diplomatic, naval, and army messages. Also, after a message had been decoded, it had to be translated before it could be read by most English-speaking officials. Some signals intercepted by British listening stations were sent to the UK for GC&CS cryptologists to attempt to decipher, and in 1939 John Tiltman managed to decode the then-current Japanese naval code.

Before Pearl Harbour, American, British, Australian, and Dutch cryptanalysts were listening to and trying to decode Japanese naval and other signals. The British office was originally in Hong Kong but then moved to Singapore; Colombo (in Ceylon, now Sri Lanka); and Mombasa, Kenya. Wireless reception there was particularly poor, and in 1943 it returned to Colombo. Following the Japanese invasion of Malaya, Army and RAF decoders were based in Delhi, India.

The principal Japanese encoding system for diplomatic messages was called Purple by American cryptanalysts, while their decoding work was known as Magic. American and British officials had managed to decode some signals before Pearl Harbour. Later they were able to decipher most Japanese messages. The Russians also succeeded in breaking into the Purple system in late 1941; the messages revealing that Japan was only going to attack the US and UK territories allowed Joseph Stalin to move considerable forces from the Far East just in time to help stop the final German push to Moscow.

Covert Communication Centres

Military establishments and training centres of all kinds were established throughout the United Kingdom, and many isolated country houses were taken over and adapted for specific purposes – not only Bletchley Park, but also Beaulieu in Hampshire and Arisaig House in Invernessshire for the training of Special Operations Executive (SOE) operatives.

Communications with forces operating overseas and Allied governments were either by radio or submarine cable. There were major pre-war radio transmitting stations at sites such as the beam wireless transmitter at Dorchester, Dorset; the very low frequency transmitter at Rugby, Warwickshire; and the wartime reserve station at Criggion, Powys. With their masts up to 250 metres in height they were impossible to conceal.

On the other hand, major efforts were made to conceal the activities taking place in the many villages and houses throughout Britain where covert communication centres were sited. One was at Porthcurno, a small coastal village near Land's End in Cornwall, which as early as 1870 had been chosen as the British terminal for the first submarine cable. By the outbreak of the Second World War it operated as many as 14 cables and was able to receive and transmit up to two million words a day – for a time it was the largest submarine cable station in the world. Situated some 150 km from Brest, it was considered to be extremely vulnerable following the fall of France, and elaborate measures were taken to protect the station's equipment. Local tin miners were engaged in June 1940 to dig two parallel tunnels, and two smaller cross-tunnels, in the granite headland behind the station. This was completed within 12 months, and the receiving and transmitting equipment was then moved into the main tunnels. The entrances, emergency exits, and operatives' housing were all camouflaged to blend in with the local topography. The station was closed in the 1970s and is now a museum

Another was at Heighton Hill, just north of Newhaven, East Sussex, and known as HMS *Forward*. In 1941 a labyrinth of tunnels was constructed in the chalk of the South Downs to house facilities for a Royal Naval headquarters. These were demolished at the end of the war, and virtually nothing was known about HMS *Forward* until Geoffrey Ellis – a Post Office engineer who as a local boy had seen it being built and was intrigued to know what had taken place there – investigated it in his retirement. His researches established that there had been two telephone exchanges, ten teleprinters, two Typex machines, a wireless telegraphy office with 11 radios, and a voice-frequency telegraph terminal for 36 channels. The tunnels housed a standby generator, an air conditioning system with gas filters, a galley, toilets, cabins for split shifts, and the recently invented phenomenon of 'daylight' fluorescent lighting. The complex was equipped for every contingency from failure of the public utilities to direct enemy action. Ten coastal radar stations along the Sussex coast reported directly to HMS *Forward* every 20 minutes. All their information was filtered and plotted before being relayed by teleprinter to similar plots at Dover and Portsmouth. The HMS *Forward* plot maintained a comprehensive maritime surveillance of everything that moved on, under, or over the English Channel from Dungeness to Selsey Bill. Further intelligence was obtained from military airfields by private telephone lines. For operational security reasons, each plot understudied its neighbour, with HMS *Forward* standing in for Fort Southwick at Portsmouth and vice versa. Wrens operated the station on a continuous three-watch rota and were supplemented by RN ratings for special occasions. On D-Day, they were joined by RAF, WAAF, and ATS personnel. HMS *Forwar*d was heavily involved in the saga of the German battle cruisers *Scharnhorst*, *Gneisenau*, and *Prinz Eugen* in February 1942; the Dieppe raid; the nightly naval Motor Torpedo Boat (MTB) harassment raids on enemy harbours and waters; the frequent SAS commando 'snoops' on the occupied French coast; the D-Day landings; and, ultimately, the liberation of France. The centre was abandoned and neglected after the war, but on Ellis' initiative the Newhaven Historical Society undertook a detailed survey of the complex, including photographs and videos. A model of the station was then produced which is on display at the Newhaven Local and Maritime Museum

American service personnel were a common sight in Britain during the latter years of the war, particularly airmen in East Anglia, soldiers in southern England, and sailors in the major ports. However, sailors were also a common sight in the small rural village of Hurley, by the Thames in Berkshire. Chosen because of its good reception for radio signals, it was Station Victor, the main UK communication centre of the US Office of Strategic Services (OSS), the predecessor of today's CIA; it was used for communicating with OSS agents on the European mainland. Operating in great secrecy, the sailors claimed to be in training but were in fact skilled radio operators – local inhabitants thought the station's transmitting and receiving towers to be elaborate radar equipment. As with HMS *Forward,* there is little in the official record about the station's activities, and it came to light only when a local resident who had been told by his father-in-law about the American sailors in the village during the war decided to investigate its history.

Underground Hideaways

In addition to the numerous secret bunkers built for the Auxiliary Units in the counties deemed to be at risk of invasion (see Chapter 3), there was also a network of bunkers with hidden radio and other communications. Comprising a relatively small number of IN and OUT stations, these would have maintained two-way contact between the Units and higher authorities had the Germans invaded. Comparatively little was known about them until recently, when a station built under a tennis court in Norfolk was accidentally discovered. This was found to include a radio room, aerials on a nearby pine tree, a cast iron pipe (which researchers say may have been used by civilian informants to drop split tennis balls containing secret messages into the bunker), an escape tunnel, and a water tank.

Decoding of enemy signals was said to have shortened the war by two years…

BRITAIN'S TIMELINE TO DUNKIRK

1939

1 Sep Germany invades Poland.

3 Sep Britain and France declare war on Germany.
Phoney War (on land and in the air) begins.
Churchill appointed First Lord of the Admiralty.
Allies start naval blockade of Germany and mine approaches to English Channel and German ports.
German submarines (U-boats) and ocean raiders start attacking Allied shipping.
SS *Athenia* torpedoed off Ireland with loss of 112 lives (28 American).
Germans start laying magnetic mines off UK east coast resulting in many sinkings.

10 Sep Britain starts sending Army and RAF units to France.

17 Sep Aircraft carrier HMS *Courageous* torpedoed and sunk off Ireland.

15 Oct First World War battleship HMS *Royal Oak* torpedoed and sunk inside RN base at Scapa Flow; Britain then starts strengthening Scapa Flow's defences.

22 Nov RN engineers recover and start to dismantle magnetic mine.

23 Nov HMS *Rawalpimdi* (armed merchant cruiser) sunk by German battlecruisers *Scharnhorst* and *Gneisenau* who then abort planned breakout into Atlantic.

13 Dec Battle of the River Plate when cruisers HMS *Exeter* and HMS *Ajax* and HMNZS *Achilles* engage the German pocket battleship *Graf Spee.* All four ships are damaged and *Graf Spee* retires to Montevideo harbour for repairs where it is eventually scuttled (a case where new German technology was outfought by superior tactics and traditional RN determination to engage the enemy more closely).

1940

8 Apr Britain lays mines off the Norwegian coast (to force shipping onto the high seas where it could be inspected by the RN).

9 Apr Germany invades Denmark and Norway.

Apr Naval engagements off Norway resulting in RN losing one carrier, two cruisers, seven destroyers, and four submarines and Germany losing three cruisers, ten destroyers, and three submarines.

11 Apr Allied troops land in Norway.

6 May Allied troops evacuated from central Norway.

9 June Last Allied troops evacuated from northern Norway.

10 May Germany invades Belgium, Holland, and Luxembourg thus outflanking the French *Maginot line* and enabling them to attack France.
British and French troops enter Belgium to engage German forces.
Churchill succeeds Chamberlain as British Prime Minister.

26 May–3 June 338,000
British, French, and Belgian troops evacuated from Dunkirk – all their equipment destroyed or abandoned.

31 May British Cabinet decides to fight on, a decision supported by the British population.

22 June Franco-German Armistice signed.

Chapter 3

LAND BATTLES LEAD TO DUNKIRK EVACUATION

But Britain wins naval victory

The disastrous first nine months of the war culminated in the heroic albeit desperate rescue of British troops from the beaches of Dunkirk at a time when early defeat stared Britain in the face. Within weeks of invading Denmark and Norway in April 1939, the Germans had invaded Holland, Belgium, and France with their new *Blitzkrieg* tactics – lightning-quick strikes with efficient communications so as to direct and coordinate its modern air force and mechanised army units. Attacking through Belgium and Holland, thus bypassing the *Maginot Line*, they achieved the overwhelming victory that led to the Dunkirk evacuation of more 300,000 British and French troops – and the French capitulation in June.

This was in fact the second humiliating evacuation within weeks, the inglorious Norwegian Campaign (April–June 1940) having ended with our defeated troops being picked up from central Norway in early May. Gordon Corrigan wrote: 'If the history faculties at universities were to run a module on how not to conduct a military campaign … then they need look no farther than the Norwegian campaign of 1940, for as complete and utter cock-ups it would be difficult to better.'

During the retreat to Dunkirk the Royal Engineers (RE) demolished some 600 bridges, blew countless craters, and built improvised jetties on the beaches to facilitate the evacuation. The Light Aid Detachments of the Royal Army Ordnance Corps

Improvised jetty made by RE from abandoned lorries photographed by the Germans. IWM

(the predecessor to the Royal Electrical and Mechanical Engineers (REME)) made valiant efforts to recover and repair damaged ordnance and vehicles, thereby enabling many items to reach the coast. Similarly, the Corps' main workshop was successfully withdrawn to the coast by its commanding officer, Lieutenant John Nicholson RAOC, who was awarded the MC for achieving this with the minimum of losses. Sadly, they then all had to be destroyed or abandoned.

In the months prior to the retreat the British Army had built up its resources and equipment and generally prepared itself for the impending battle. Royal Signals, for instance, made use of the French civilian landline network with Army personnel working alongside local staff in their exchanges. Wireless telegraphy was regarded as a back-up to the landlines, the network being reinforced by the laying of additional dedicated lines where necessary.

The hard lessons learned during the disastrous Norwegian and French campaigns made it immediately apparent that the procedures envisaged pre-war would have to be drastically rethought and that new techniques would need to be developed and applied. Principal among these operations were the importance of airpower, tactical mobility, close inter-service cooperation, rapid decision-making – and the need for modern equipment.

Defeating the Magnetic Mine

As Britain's fate hung in the balance, a crucial engineering challenge was presented to Allied scientists: how to counter the menace of the mines sown on the bed of Britain's shallow coastal waters by German ships and submarines or parachuted there by the *Luftwaffe*. So serious was the problem that the Port of London, then Britain's busiest port, was nearly closed.

These mines were detonated by the magnetic field of a ship passing over them, and it is notable that before those 800 'little ships' sailed from Kent ports to rescue the troops stranded

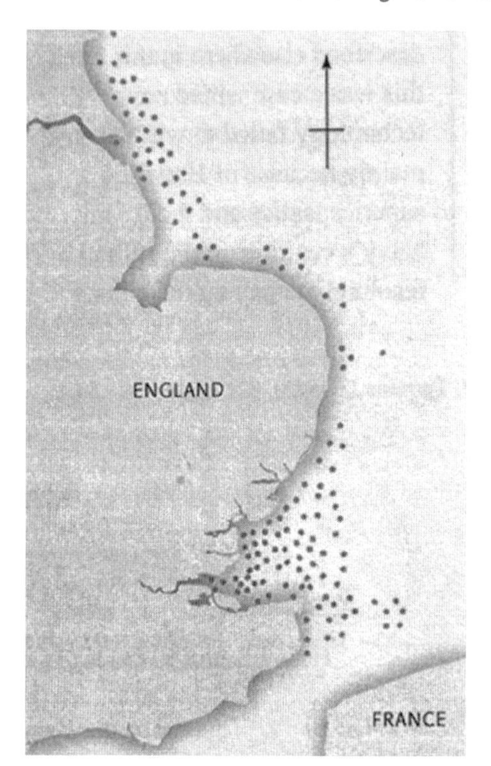

Sinkings off UK's east coast September 1939–December 1940. Chatham Historic Dockyard

HMS Belfast. *Now moored near Tower Bridge. IWM*

in Dunkirk, they had to be 'wiped' by teams from HMS *Vernon* to make them magnetically neutral.

The importance of the ensuing engineering feat was later expressed by one of its heroes, Commander Sir Charles Goodeve: 'Although in the technical achievement the human effort was not in the same class as the radar or U-boat battle, it was the first technical battle in which we won a decisive victory over the enemy; but more important still, it was one which brought science fully into the war in the very early days.'

The earliest casualty of the magnetic mines, the SS *City of Paris,* was damaged on 16 September while among naval casualties before the end of the year were the battleship HMS *Nelson,* which was put out of service for nearly a year; the newly commissioned cruiser HMS *Belfast,* which broke its back and took two years to be repaired; the destroyer HMS *Blanche,* which was sunk; Captain Lord Louis Mountbatten's ship, the destroyer HMS *Kelly,* which had her stern blown off; and the minelayer HMS *Adventure,* which suffered serious damage.

Caught unawares, and ignorant of how they operated, the Navy's experts had to wait for mines to be dropped until they could be recovered and dismantled before they exploded. Two such mines were dropped by the *Luftwaffe* on the mud flats at Shoeburyness, Essex, in November 1939, and defused and dismantled by Lt Cdrs John Ouvry and Roger Lewis RN, and by CPOs Charles Baldwin and Archie Vearncombe of HMS *Vernon.* The pairs were awarded DSOs and DSMs respectively for their bravery.

Thereafter, it was generally possible for the mines that could be found to be recovered and made safe, although sometimes things went wrong, with tragic results for those attempting to defuse them. The Germans also sometimes used delayed-action mines or booby-trapped the fuse mechanisms, adding to the hazards faced by the naval defusing teams. Among naval officers engaged in such work was Sub-Lieutenant (Sp) Peter Danckwerts RNVR, who was awarded the George Cross for his bravery. Detailed

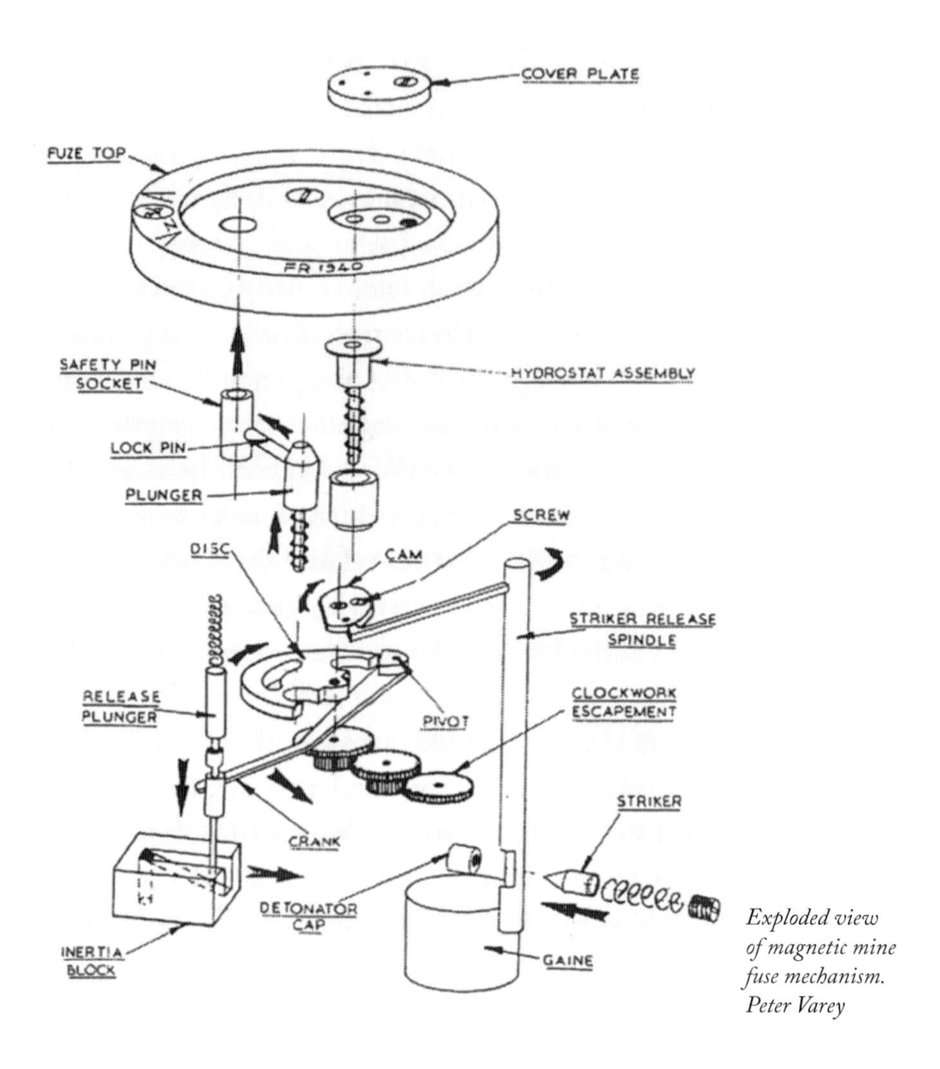

Exploded view of magnetic mine fuse mechanism. Peter Varey

accounts of the mechanisms involved (which he described as 'a miracle of ingenuity') are contained in Danckwerts' biography.

Solutions also had to be found to protect ships from the mines that had been successfully dropped on the sea bed. A variety of counter-measures were investigated as a matter of urgency, but the two systems developed by Lieutenant Commander Charles Goodeve FRS RNVR proved most effective and were widely adopted. The first was to develop a system of detonating the mines which did not damage the vessels involved, and the second was to demagnetise ships.

The former aim was achieved by the Double L Sweep, which involved two small wooden-hulled minesweepers towing a long loop-shaped buoyant electric cable between and behind them, through which strong electric DC currents were passed; the magnetic fields then induced in the sea were sufficient to detonate the mines.

Danckwerts, Professor Peter GC MBE FRS (1916–84)

The eldest son of Vice-Admiral Victor Danckwerts, who had seen action in HMS *Kent* during the 1914 Battle of the Falkland Islands and had been the Navy's director of plans in 1940. He fell foul of Churchill when the latter was First Lord of the Admiralty and was removed from office, but later served in Washington before becoming second-in-command of the Eastern Fleet.

After schooling at Winchester, Peter Danckwerts read chemistry at Oxford before joining the Royal Navy in 1940. Having volunteered for special duties, he was assigned to undertake bomb and mine disposal.

After a short training period, S/Lt (Sp) Danckwerts RNVR led a small team defusing magnetic and other mines dropped in the Thames Estuary between Teddington and Southend. He was awarded the George Cross later that year for 'great gallantry and undaunted devotion to duty' whilst neutralising enemy mines.

In 1942 he was posted to Gibraltar to deal with the threat of Italian mines, including limpet mines attached to ships by frogmen riding midget submarines. Then, after a spell in Algiers, he participated in the invasion of Sicily but, ironically, he had the misfortune to tread on a small anti-personnel mine. After hospitalisation in Portsmouth, he spent the last year of the war with the planning staff at the headquarters of Combined Operations.

Following demobilisation, he studied chemical engineering at MIT and then worked as an academic in Britain before joining the UK Atomic Energy Authority. He was appointed professor of chemical engineering at Imperial College and then at the University of Cambridge. He was elected president of the Institution of Chemical Engineers in 1965 and a Fellow of the Royal Society in 1969. After retiring from the University of Cambridge in 1976 he became the executive editor of *Chemical Engineering Science.*

He died in Cambridge on 25 October 1984.

The demagnetising of metal ships was accomplished by degaussing the hulls of the ships. Initially this was done through a process known as coiling, which involved wrapping electromagnetic coils around the hulls and passing strong electrical pulses through them, thereby making the ships magnetically neutral. This was too expensive to be used universally, but Goodeve devised a cheaper system called wiping, which involved dragging a large electrical cable with a 2,000 amp current passing through it

Goodeve, Commander Sir Charles OBE FRS RNVR (1904–80)

Born and educated in Canada, he obtained a BSc in 1925 from the University of Manitoba after studying physics and chemistry. He served in the Winnipeg division of the Royal Canadian Naval Volunteer Reserve before moving to University College London in 1927. Here he later became a reader in physical chemistry, his research work there resulting in his election as a Fellow of the Royal Society in March 1940.

After transferring to the RNVR he attained the rank of lieutenant-commander RNVR and was attached to HMS *Vernon*, the RN's torpedo and anti-submarine headquarters which also oversaw electrical developments. Having successfully developed countermeasures for magnetic mines, in the summer of 1940 he became the senior technical officer in the Admiralty's newly established Department of Anti-Aircraft Weapon Development, later renamed the Department of Miscellaneous Weapon Development, (DMWD), which was also known as the 'Wheezers and Dodgers'. Subsequently promoted to commander he was appointed the Admiralty's Assistant Controller, Research and Development, in 1942 with broad oversight of the Royal Navy's research and development work.

He was instrumental in developing the 'Double L Sweep' for exploding magnetic mines safely and the degaussing system for making ships magnetically neutral (for which he received £7,500 from the Royal Commission on Awards to Inventors), and the Hedgehog depth charge launcher that was credited with the destruction of some 50 U-boats.

He was knighted at the end of the war and also awarded the US Medal of Freedom.

Post-war he served as director of the British Iron and Steel Research Association where, among other things, he introduced the principles of Operational Research. Following retirement in 1969, despite failing health due to the onset of Parkinson's disease, he immersed himself in various activities including the establishment of the Operational Research Society. He died in 1980.

along the side of a ship. Goodeve was one of a number of scientists who received awards for the development of the degaussing technique, his prize being £7,500.

Had these scientists and naval officers not found solutions as rapidly as they did, the early months of the sea war would have proved to be extremely costly as well as dangerous for Britain. Interestingly, the same technology is still largely applicable today.

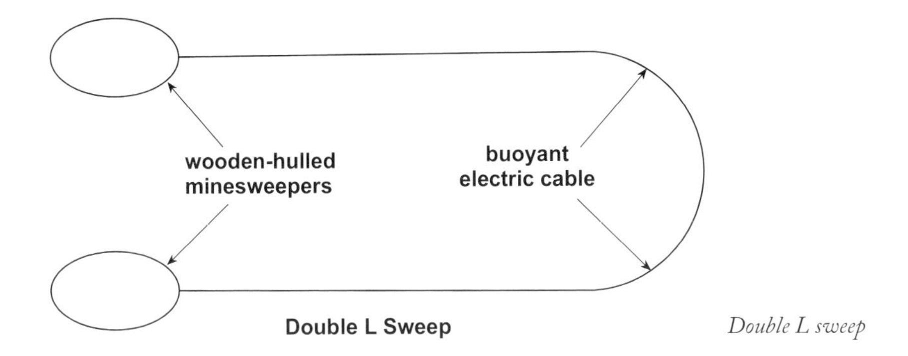

Double L sweep

Flying Baskets

The early days of the war were as harrowing in the air as on land and at sea. One air raid in late December 1939 was described in the Official History as a 'disaster', with over 50% of the aircraft lost, mostly to enemy fighters – an example of the difference between British and German aerial capabilities at that time. Lessons were learned, however, including the need for better radio communications and the recognition that losses incurred in night attacks were significantly lower than those in daylight raids.

From the start of the war the RAF had also been engaged in dropping propaganda leaflets over major German cities. While the RAF hierarchy correctly considered this to be of minimal value, they recognised that there were considerable training benefits.

Wellington bomber fuselages prior to covering with waterproof fabric. IWM

Wallis, Sir Barnes
CBE FRS FRAeS MICE (1887–1979)

After serving an apprenticeship and acquiring an external engineering degree, he joined Vickers in 1913, with whom he continued until his retirement in 1971. Working on airship design he created the geodetic airframe (comprising a spiral crossing basket-weave of load-bearing members) which was used on the Airship R100 and then on the Wellington bomber.

Recognising the potential of selective as well as saturation bombing, he designed the bouncing bomb, (used to attack the Ruhr Valley dams), and deep-penetration bombs. These weighed up to 10 tons each and were used against U-boat pens, the Tirpitz battleship, and V2 rocket launch sites.

Post-war he led Vickers' R&D division where he investigated supersonic flight and swing-wing technology. He also developed an experimental rocket-propelled torpedo. In 1955 he was appointed consultant to the project to build the Parkes radio telescope in Australia and five years later proposed developing large cargo submarines for the transport of goods.

Due to the high casualty rate in the bomber crews who attacked the Ruhr Valley dams, he later strove hard to reduce the risks for test pilots, extensively testing his designs on models and becoming a pioneer of the remote control of aircraft. Awarded £10,000 for his war work, he donated it all to Christ's Hospital for the benefit of the children of RAF personnel who had been killed or injured.

He was still full of new ideas in his retirement – and bemoaned the fact that the aircraft designers of the day would not take them up! He was elected a Fellow of the Royal Society in 1945 and knighted in 1968. He died in 1979 and is buried in Effingham, Surrey, where his headstone carries the Latin inscription *Spernit Humum Fugiente Penna* ('severed from the earth with fleeting wing').

Crews were also instructed to observe anything of interest that could be seen on the ground below.

Among the bombers used in this period was the Wellington, whose airframe was designed by Barnes Wallis on the geodetic principle – a basket-weave of diagonally crossing load-bearing members. As such it had significantly more strength that conventional airframes, and there were many examples of seriously damaged

Wellington aircraft returning to their bases. The downside of the technique was that the planes were more complicated to build, and it was not repeated in later aircraft.

But Britain now had to use all its ingenuity and resources to defend her island home and to ensure her survival…

RADAR, BATTLES IN THE AIR, AND THE *BLITZ* TIMELINE

Radar

1935	Radio Research Station (RRS) scientists detect radio waves reflected from an aircraft 25 km distant. Government authorises further investigations.
1936	RRS moves to country house near Felixstowe, Essex.
1939	Chain of radar transmitting and receiving towers around the coast from the Solent to the Tay Estuary operational. Dowding system for interpreting information received and controlling aircraft movements also operational.
1940	Chain extended to cover southwestern and western coastlines.

Battle of Britain (BoB)

1933	Air Ministry issues specification for new fighter aircraft. Designing Supermarine Spitfire and Hawker Hurricane fighters gets underway. Rolls-Royce conduct first test run of Merlin aero-engine.
1937–38	First deliveries of Hurricanes and Spitfires to RAF.
1940	Dowding commander in chief of Fighter Command and Park in charge of No. 11 Group covering South East England.
10 Jul	BoB commences with *Luftwaffe* sweeping over SE England but with RAF conserving resources.
1 Aug	Hitler orders *Luftwaffe* 'to destroy the British air force'.
8 Aug	*Luftwaffe* begins attacks on RAF bases. RAF engages enemy.
15 Aug	Main *Luftwaffe* attack repulsed by RAF. Operation Sealion for invasion of UK postponed.
24 Aug	Start of daytime raids on London
31 Oct	BoB ends.
Nov, Dec	Dowding succeeded by Sholto Douglas and Park by Leigh-Mallory.

THE BLITZ

7 Sep	Day and night raids begin on London and South East England and continue for the next 57 days.
1 Nov	*Luftwaffe* initiates concentrated night raids on London, ports and industrial targets (the *Blitz*).
14 Nov	Major raid on Coventry.
1941	Night fighters fitted with AI radar.
1–7 May	Major raids on Liverpool and Merseyside.
May	*Blitz* ends as *Luftwaffe* units withdrawn in readiness for invasion of Russia.

Chapter 4

RADAR, BATTLES IN THE AIR, AND THE *BLITZ*

Victories followed by survival

Winston Churchill memorably eulogised the brave pilots who fought the Battle of Britain, and so saved the nation, in those crucial make-or-break months from July to October 1940: 'Never in the field of human conflict was so much owed by so many to so few.'

The Prime Minister well knew, however, that it was an invisible but immensely powerful engineering development, which was no less significant than those celebrated dog-fight heroics, that had made the epic victory possible.

'The RAF's Hurricanes and Spitfires were handicapped by clumsy tactical doctrine and .303 machine-gun armament with inadequate destructive power,' Max Hastings wrote in *All Hell Let Loose*, 'but squadrons were controlled by the most sophisticated radar, ground-observer and voice-radio network in the world, created by an inspired group of civil servants, scientists and airmen.'

With radar, he added, 'the RAF had developed a remarkable system of defence, while their opponents had no credible system of attack.'

That alleged German weakness would not have been apparent to most Britons in the days and months after the Dunkirk evacuation. By mid-1940 the British and French armies had been defeated in Norway and France, and Norway, Denmark, the Low Countries and the northern and western coastal areas of France had been occupied. Therefore, Germany was able to base her aircraft, U-boats, and surface vessels along the whole continental coastline from northern Norway to the Franco-Spanish border.

Britain had thus became vulnerable to short-range aerial attack and possible invasion while the German Navy had easy access to the North Atlantic and could more easily attack the merchant vessels supplying Britain with her essential imports of food, fuel and armaments.

The country was in the gravest peril, though the population was resolutely behind the Government and faced the prospect with equanimity. Their attitude was well summarised by the reputed remarks of a West End club porter who said, 'Well, sir, we've made the final and, what's more, we are playing at home!'

Proving Baldwin Wrong

During the inter-war years the conventional wisdom, forcibly expressed by the then-Prime Minister, Stanley Baldwin, had been that there was no defence against aerial attack other than having a greater force than the enemy.

'It is well also for the man in the street to realise,' he told Parliament in 1932, 'that there is no power on earth that can prevent him from being bombed. Whatever people may tell him, the bomber will always get through. The only defence is in offence, which means that you have to kill more women and children more quickly than the enemy if you want to save yourselves.'

Enter the engineers!

While the RAF concentrated most of its research and development efforts on Bomber Command, to the detriment of aerial defence and Fighter Command (and, to an even greater extent, maritime warfare and Coastal Command), brains were already at work on the kind of defensive shield Baldwin had been unable to imagine.

It was thought possible, for example, that the noise made by aircraft might be used to detect their direction and range, and experimental acoustic mirrors were erected around the coast, some of which are still standing. Measuring up to 70 metres wide and 5 metres high, they were made of concrete to spherical, not parabolic, concave shapes and had a microphone near the focal points which could be moved so as to determine the aircraft's direction. While partially successful, they became less effective with increasing aircraft speed and were abandoned when electronic systems were developed.

Following discussions between Sir Henry Tizard, the chairman of the Aeronautical Research Committee, and Robert Watson-Watt, the superintendent of the radio research station then based at Slough, Berkshire, Watson-Watt investigated the feasibility of developing and using damaging radiation – commonly known as 'death rays' – as a defence. He concluded that they would not damage the crew or equipment inside a metal aircraft, which would in effect act as a Faraday cage (a mesh of conductive material around a person or object which shields the contents from external electromagnetic fields).

However, his assistant, Arnold (Skip) Wilkins, suggested that it might be possible to detect

Acoustic mirrors, Denge, Kent.
© Paul Russon

Wilkins, Arnold ('Skip') OBE (1907–1985)

Educated at the universities of Manchester and Cambridge, Wilkins was a member of Watson-Watt's team. His suggestion that it might be possible to bounce radio waves off an aircraft led to the 1935 experiment where a reflected wave was detected in a field near Daventry. Later that year he moved to Orford Ness in Suffolk to conduct further experiments which eventually led to the move of the Telecommunications Research Establishment there. He was then instrumental in setting up the coastal radar network before helping to develop the British version of the Identification Friend or Foe (IFF) system.

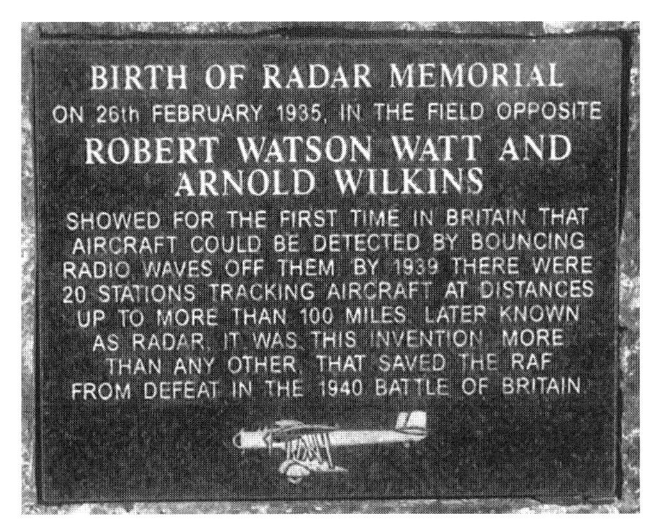

Commemorative Plaque, Stowe Nine Churches, Northamptonshire. Peter Mallows

an aircraft by bouncing radio waves off it. Following experiments using radio waves generated by the BBC shortwave transmitter at Daventry, in February 1935 they were able to detect an RAF bomber at a distance of 12 km. This was the first time that this had been achieved in Britain and the event is now commemorated by a plaque at Stowe Nine Churches, Northamptonshire.

In consequence, in March 1935, Tizard's committee recommended that the Treasury authorise £10,000 for large-scale experiments to be carried out under Watson-Watt's leadership. By January 1936, these had demonstrated that it was possible to detect a plane 40 km away. New premises were acquired at Bawdsey Manor near Felixstowe, Essex, which had extensive grounds, to further investigate and develop

Coastal radar station, Sussex, 1940. Left: 110 metre steel transmitting towers. Right: 72 metre wooden receiving towers. Science at War (HMSO)

Radar transmitting and receiving station at coastal radar station. Science at War (HMSO)

the Radio Direction Finding (RDF) technology, which was later named radar (**RA**dio **D**etection **A**nd **R**anging). Following further successful investigations, the Treasury authorised £10 million for 20 stations to be built around the coast from the Firth of Tay to the Solent, each comprising a number of steel transmitting towers and lower wooden receiving towers.

In the development of radar the scientists worked closely with Air Marshal Sir Hugh Dowding, the Air Member for Supply and Research in 1930–36 and later Fighter Command's commander in chief. While Harry Wimperis, an aeronautical engineer who was the Director of the Air Ministry's Department of Scientific Research, fought inter-service battles, Watson-Watt recruited and led the research team; foresaw both the possibilities and the problems; and fought the team's administrative battles. Among key team members were Jimmy Rowe, A. F. Wilkins, and Dr E. G. 'Taffy' Bowen (later instrumental in developing airborne radar), while committee members who gave influential support included the Nobel Prize winner Professor Archibald Hill and Professor Patrick Blackett. Co-operation between scientists and serving officers was exemplary; Tizard described it as 'the great lesson of the last war'.

By the summer of 1939 the coastal radar chain comprising separate high- and low-level systems (known as Chain Home and Chain Home Low respectively) was operational, with 21 and 30 stations respectively. The fall of France not having been anticipated, the original chains had to be hastily reconfigured and augmented in the summer of 1940 to provide cover west of the Solent and for ports and cities in Wales and the North West.

The Observer (later Royal Observer) Corps, which had been established pre-war, supplemented the information derived from the radar network by reporting numbers, direction, height, and composition of enemy formations.

Despite the high visibility of the radar towers, the Germans amazingly did not make them prime targets, nor were they mentioned in a July 1940 report by the *Luftwaffe*'s Chief of Intelligence on the British defence system.

Rowe, Albert ('Jimmy') CBE (1898–1976)

A government scientist, Rowe was secretary of Tizard's committee that oversaw the research that led to the establishment of radar. He succeeded Watson-Watt as superintendent of the Folkestone Telecommunications Research Establishment and gave priority to the erection of the coastal radar station chain. Following the Battle of Britain he directed research which led to the development of airborne radar and centimetric radar with the cavity magnetron. Despite some opposition from Bomber Command, who felt that the projects would not produce large-scale results, Rowe also led in the development of the Oboe navigation system and of H2S radar. In 1946 he moved to Australia as chief scientific officer for the British Rocket Programme. The following year he joined the Australian Department of Defence before serving as vice-chancellor of the University of Adelaide in 1948–58. He is credited with coining the phrase 'Operational Research' in 1935. He suffered from poor health and died in 1976.

Watson-Watt, Sir Robert KCB FRS FRAeS (1892–1973)

A government scientist who joined the Royal Aircraft Factory at Farnborough in 1914, he first worked on radio systems for detecting thunderstorms. By 1933 he was superintendent of the Radio Research Station at Slough and then, from 1936 of the RAF's Radio Direction Finding research station near Folkestone. Having concluded that a 'death ray' of intense radio beams could not bring down a plane, he and his team realised that radio waves were reflected from aircraft and could be detected. Thereafter he designed the radar network that was installed and fully operational by the start of the war (and for which he was awarded £52,000). During the war he became the Air Ministry's scientific advisor on telecommunications. Post-war he led British delegations to international conferences, established a consultancy, and lived for a while in Canada. He served as president of the Royal Meteorological Society, the Institute of Navigation, and the Institute of Professional Civil Servants, and as vice-president of the Institute of Radio Engineers in New York. Reputedly, in retirement he was caught for speeding by a radar speed-trap!

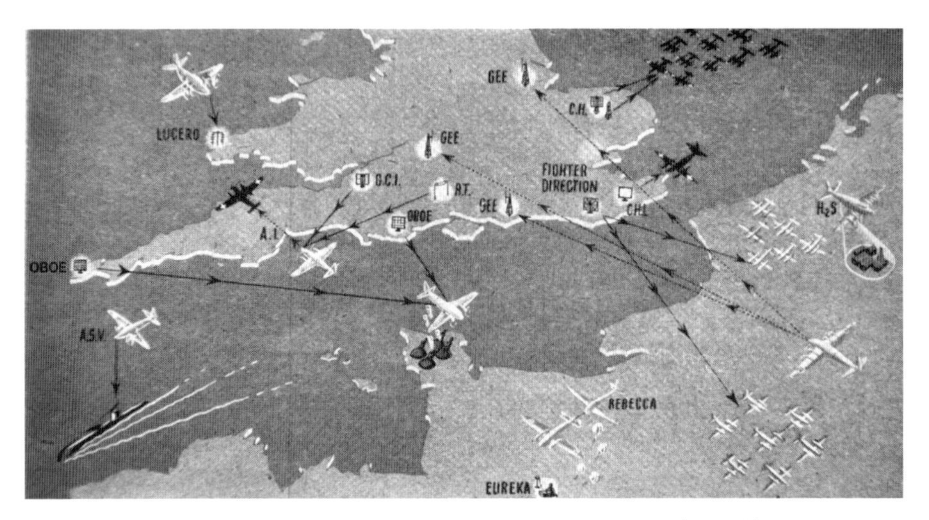

Radar variants developed by Allied scientists and engineers. Science at War (HMSO)

Other radar systems which later evolved from the Bawdsey team's pioneering work included Airborne Interception (AI) radar (for night fighters); Air-to-Surface Vessel (ASV) radar (to aid attacks on ships); Oboe system (for improved bombing accuracy); Gee system (also for improved bombing accuracy as well as assisting bombers return to their bases); H2S (target finding for bombers); and Rebecca and Eureka used by invading paratroops. The Telecommunications Research Establishment and the Admiralty Signals Establishment were both in the forefront of such developments.

The Royal Commission on Awards to Inventors gave a tax-free sum of £87,950 to the Bawdsey team, Watson-Watt's share being £50,000 (award No. 171). The transition from the flickering and truant radio echoes of 1936 into the reliable defence system of 1940 was 'one of the greatest combined feats of science, engineering and organization in the annals of human achievement'.

Dowding Joins up the Dots

Following the development of radar, Air Chief Marshall Hugh 'Stuffy' Dowding designed an integrated air defence system of Ground-Controlled Interception (GCI). Known as the Dowding system, it constituted: a number of radar stations; the Observer Corps', human observers; raid plotters at Fighter Command's headquarters at Bentley Priory (a converted country house near Stanmore, north London), and at various Group headquarters; and control of aircraft by radio. The land-based components were integrated by dedicated phone and teleprinter links buried sufficiently deep enough to provide protection against bombing.

In the operations room at Bentley Priory sat Dowding and Air Vice Marshal Keith (later Sir Keith) Park, a New Zealander who was the commander of No. 11 Group and

Dowding, Air Chief Marshal Lord Hugh ('Stuffy') GCB GCVO CMG (1882–1970)

An austere character, Dowding was originally a Royal Artillery officer; he joined the Army in 1899 and in 1912 learned to fly at his own expense. He transferred to the RFC in 1914 and was for a period in the First World War in command of a wireless experimental establishment at Brooklands. Lucky to escape compulsory retirement at the end of the war, he then commanded RAF stations in London and Baghdad until 1930, when he was appointed to the Air Council as Air Member for Supply and Research. In this capacity he oversaw the introduction of a new generation of fighter and bomber aircraft and worked closely with Watson-Watt in establishing the radar network. In 1936 he was appointed commander in chief of the newly formed Fighter Command and established the integrated Ground Controlled Interception (GCI) system that operated the fighter aircraft during the Battle of Britain. He was then already past retirement age; despite the battle's victorious conclusion, he was removed from office in late 1940 prior to retiring in 1942 after heading the Air Mission to the USA. His statue was erected outside the RAF's church, St Clement Danes, in the Strand, London.

as such responsible for Fighter Command's operations in southern England. Also there were Lieutenant-General Sir Frederick Pile, commander of the Army's Anti-Aircraft Command; the commandant of the Observer Corps; and liaison officers from the RAF's Bomber and Coastal Commands, the Royal Navy, Ministry of Home Security, and the Civil Defence Association. The number of German planes brought down by anti-aircraft fire was small, however, compared to those shot down by the RAF.

Dowding and Park (and they alone) received copies of German signals that had been decrypted at Bletchley Park, initially via the Air Ministry but later directly by dedicated teleprinter. The information was filtered and then disseminated to Group 11 headquarters at Stanmore, north London, and to other group headquarters when necessary, and thence to sector operations rooms, the final link in the command chain. The sector controllers were the key men who bore the great responsibility of putting the squadrons in the air, positioning them, having executive authority over them until they saw the enemy, and eventually guiding them back to their bases. These controllers needed 'a sense of judgement amounting to an intuition', and they regarded themselves as much servants as commanders of the pilots they controlled.

This integrated monitoring, communications, and control system was essential in enabling RAF fighters to be airborne and in the right place and height to attack the incoming bombers.

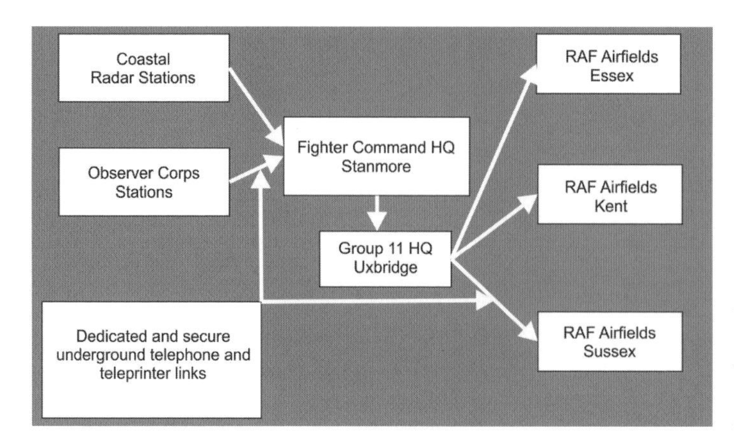

Dowding GCI System of Monitoring, Communications, and Control

Meanwhile, since October 1936 a small group of army scientists had been attached to Watson-Watt's radar establishment. Over the following years they used radar-derived techniques to develop means of improving the accuracy of gun laying (GL) for anti-aircraft guns. The original techniques using 6 metre GL sets gave indifferent results, although the arrival of centimetric radar improved performance fivefold. Shooting down a raider in 1940, for example, on average required firing 20,000 rounds, but this was reduced to 4,000 rounds by the following spring. Proximity fuses were also created and proved to be very powerful anti-aircraft devices; they were likewise designed for use in shells, rockets and bombs. Details of their development were among the scientific secrets shared with the Americans by the Tizard mission in 1940.

As for the Germans, they were later discovered to have been developing their own '*Freya*' radar system, though it was too late for the Battle of Britain. Indeed, Major (later General) Adolf Galland, who flew Messerschmitt 109s during that conflict, wrote: 'The British had, from the first, an extraordinary advantage, never to be balanced out at any time in the whole war: their radar and fighter-control network. It was for us and for our leadership a freely expressed surprise, and at that time a very bitter one, that Britain had at its disposal a close-meshed radar system, obviously carried to the highest level of current technique, which supplied the British fighter command with the most complete basis for direction imaginable ... We had nothing like it.'

Magic Merlin

As for those Spitfires and Hurricanes, they began the war with insufficient firepower, as Hastings asserted, but they had the benefit of one superlative piece of engineering – the Merlin engine.

Following Rolls-Royce's usual practice, the engine was named after a bird of prey, in this case a small falcon. It was designed by a team working under Ernest Hives, then the company's general works manager and newly appointed board member who,

Hurricane and Spitfire Fighter Airplanes

Among the persons principally responsible for the RAF being equipped with Hurricane and Spitfire fighters were Air Chief Marshal Sir Edward Ellington, who introduced a number of far-reaching expansion schemes; Wing Commander R. T. Williams and Squadron-Leader Ralph Sorley from the Air Ministry; and Sidney Camm and Ronald Mitchell, the chief design engineers of the Hawker Aircraft Company and the Supermarine Aviation Company respectively. The 1934 Air Ministry Specification F.5/34 was the original basis for these two fighter aircraft. However, following discussions between the officers and the two manufacturers, modified specifications were developed, F.36/34 and F.10/35 being the geneses of the Hawker Hurricane and the Supermarine Spitfire fighters respectively. The specifications required aircraft with a minimum speed of 500 km/h (310 mph) and were designed to be powered by the newly developed Rolls-Royce Merlin engines.

Sorley, Air Marshal Sir Ralph KCB OBE DSC DFC FRAeS FRSA (1898–1974)

Sorley was a pilot with the Royal Naval Air Service and the Royal Air Force during the First World War. In the 1930s he was instrumental in the creation of the specifications from which evolved both the Supermarine Spitfire and the Hawker Hurricane fighter aircraft. A senior commander during the Second World War, he founded the Empire Test Pilots' School for training pilots in Canada and other overseas countries. He retired from the RAF in 1948 and joined De Havilland, where he assisted in the development of air-to-air missiles.

Camm, Sir Sydney CBE FRAeS (1893–1966)

Appointed Hawker Aircraft Company's chief designer in 1925, Camm was responsible for the design of the wartime Hurricane, Typhoon, and Tempest aircraft and the post-war Hunter and P.1127 aircraft. The latter was the progenitor of the Harrier, the world's first vertical take-off and landing (VTOL) aircraft. At the end of his career he was planning a plane to travel at Mach 4. The RAeS, of which he was president in 1954–5, has since 1971 held the biennial Sir Sydney Camm Lecture.

Mitchell, Reginald J. CBE FRAeS MICE (1895–1937)

In 1920, when only 25, he was appointed chief designer and chief engineer of the Supermarine Aviation Company, which specialised in the design and manufacture of seaplanes. These included a number of flying boats for the RAF and of high-speed racing seaplanes which were entered for the annual Schneider Trophy races held in the early inter-war years. Two Supermarine S.5 aircraft were entered in 1927, and finished first and second. The Supermarine S.6 won in 1929. The final entry in the series, the Supermarine S.6B, marked the culmination of Mitchell's quest to 'perfect the design of the racing seaplane'. The S.6B won in 1931 and broke the world air speed record 17 days later. When Vickers took over the Supermarine Company in 1928, they made it a condition that Mitchell should remain in post for at least five years. Following the issue by the Air Ministry in 1934 of the specification for a new fighter, Mitchell designed the Spitfire, bringing together a number of ideas originated by others but combining them and incorporating his own skill in designing high-speed aircraft. The first prototype was rejected because of unsatisfactory performance; however, a redesigned aircraft won the approval of the Air Ministry, and the prototype of the modified plane first flew in March 1936. During the early war years the Spitfire proved to be the most effective fighter aircraft. Altogether, Mitchell designed 24 aircraft. Sadly he died in 1937 aged 42, and thus did not live to see the vital role his aeroplane played in the defence of Britain and the free world.

The Hurricane was produced first, making its maiden flight in November 1935 and going into service in December 1937. The Spitfire was slower in reaching the squadrons, the first test flight being in March 1936 and it entered service in August 1938. During the Battle of Britain Herman Goering, who was the head of the Luftwaffe, is alleged to have said, 'Get me some Spitfires'. Altogether, 12,780 Hurricanes and some 22,000 Spitfires were eventually produced.

Sadly Williams died in 1934 and Mitchell in 1937, so neither lived to see the vital importance of their pioneering work. Sorley and Camm on the other hand each did more sterling work both during and after the war. Sorley reached the rank of Air Marshal and Camm designed many more innovative aircraft; both were knighted.

Hives, Baron (Ernest), CH MBE MIMechE (1886–1965)

Having worked as a driver and mechanic with small companies, Hives joined Rolls-Royce in 1908 to undertake experimental work and test new production vehicles. He was one of the first people to achieve 100 mph on a racing circuit. He is regarded as one of the four people – along with Rolls, Royce, and Johnson – to have established the company's reputation and its premier position in aero-engine manufacturing. During and after the First World War he designed aero-engines and led the design of the Merlin engine as a private venture by the company, initially to power the Spitfire and Hurricane fighters, but later for a variety of other aircraft. Enraged by official complacency, and in anticipation of the early outbreak of war, he prepared the company for large scale production – in the event some 112,000 engines were manufactured in the UK plus some 37,000 under licence in the USA. In 1942 he negotiated for Rolls-Royce to take over the development of Air Commodore Frank Whittle's gas turbine. He became managing director at Rolls-Royce in 1946 and chairman in 1950. He retired in 1957.

Hooker, Sir Stanley, FRS FRAeS MIMechE (1907–84)

Educated at Imperial College and Oxford, he first served at the Admiralty but moved to Rolls-Royce in 1938, where he researched superchargers on Merlin engines and was able to increase the performance of the early Spitfire engine by 30%. By mid-1942 his two-stage supecharged engine further increased performance and was fitted on all new variants. Concurrently, he was collaborating with Frank Whittle in the production of his jet engine. He was instrumental in arranging the transfer of production from Rover to Rolls-Royce in 1942 in exchange for taking over the latter's tank engine factory in Nottingham. He then collaborated with the American company General Electric in the production of jet engines. He left Rolls-Royce in 1949 to take charge of Bristol Aero Engines. After Rolls-Royce went into receivership after the RB.211 fiasco he returned there in 1971 in an advisory capacity and when the company was nationalised later that year he was appointed a main board director in charge of technology. He finally retired in 1978. He was elected FRS in 1962 and knighted in 1974.

after 1940, was ably assisted by Stanley Hooker. Both the Hurricane and Spitfire were designed around the Merlin engine.

The V12 liquid-cooled piston engine was first test run in October 1933. Originally it had an evaporative cooling system but this proved unreliable; when supplies of ethylene glycol from the US became available the engine was adapted to use a conventional liquid-cooling system. Once it was decided that the engine should power the two new fighters, its development was given top priority, production contracts were placed, and government funding provided. The demands of war brought about a number of developments to improve its performance.

Manufactured at factories at Derby, Crewe, Glasgow, and Manchester, (and later at various Packard factories in the USA), it was one of the most successful aero engines of the war and it powered, in addition to the Hurricane and Spitfire fighters, the Lancaster bomber and many other aircraft. It was also used in the American P-51 Mustang long-range fighter-bombers, thereby transforming their performance (see Appendix 5). Throughout the Battle of Britain spare parts for engines were hastily ferried to the fighter squadrons' maintenance engineers.

During the dog fights of 1940, it became apparent that the Merlin engines had a serious problem with their carburettors when manoeuvring in combat. The negative 'g' created by suddenly lowering the nose of the aircraft resulted in the engine being

Rolls Royce Merlin Engine photographed at Pearce Air Force Base Western Australia. JAW at English Wikipedia

Shilling, Beatrice OBE PhD MSc (1909–90)

A successful aeronautical engineer and motorcycle racer, Beatrice Shilling made her mark in two male-dominated environments. After an electrical engineering apprenticeship in Devon and engineering degrees at Manchester, she joined the RAE at Farnborough in 1936, where she became the leading specialist in aircraft carburettors. After the Merlin engine-powered Hurricanes and Spitfires went into service it was discovered that the engines could cut out in a steep dive. The solution was a simple but ingenious device – a small brass disc with a hole in the middle fixed in the engine's carburettor. Known as 'Tilly's orifice' it was later used on all Allied aircraft. A keen motorcyclist, she was awarded a Gold Star at Brooklands for lapping the track at over 170 km/h (106 mph) on her Manx Norton 500, thereby making her the fastest female on two wheels at the time. She worked at the RAE until her retirement in 1969.

flooded with excess fuel, causing it to lose power or shut down completely. German fighters used fuel injection engines and did not have this problem, so they could evade RAF fighters by flying negative 'g' manoeuvres that could not be easily followed. A simple yet ingenious solution was devised by Beatrice Shilling of the Royal Aircraft Establishment, Farnborough. Officially called the 'RAE restrictor', it was a small metal disc with a hole in the middle fitted into the engine's carburettor. Although not a complete solution, it allowed RAF pilots to perform quick manoeuvres without the loss of engine power. The restrictor was immensely popular with pilots, who affectionately nicknamed it the 'Tilly orifice'. It continued in use as a stopgap until the introduction of the pressure carburettor in 1943.

Down on the Ground

From the beginning the RAF was able to pick up *Luftwaffe* Morse transmissions and later, using an ever-increasing number of RAF and WAAF German linguists, it started listening in to *Luftwaffe* radio broadcasts. (This was the start of the Y-service, which monitored German radio broadcasts throughout the war.) Even early on, having gained knowledge of the German call signs and frequencies, the monitors were able to indicate the height of the incoming formations and their likely actions as well as predict the aircraft type and base from which they were operating; this was most helpful for the controllers, who could then anticipate the likely return route of the enemy aircraft. When direction-finding facilities became available, the monitors were able to determine the precise positions of the incoming formations.

Another feature of the battle was that the German aircraft had to expend a significant amount of their fuel in travelling to South East England, thus restricting the time they were able to spend on their intended mission. Also, any damaged planes that were able to land were lost to the Germans. The British planes, on the other hand, could use all their fuel in action, and if they had to make a forced landing, they were on friendly territory. Moreover, both pilots and planes could frequently be back in action within a short period.

Quick servicing of the aircraft by RAF engineers and other ground staff was vital, and it had been developed to a fine art. Refuelling, rearming, engine checking (including oil and glycol coolant), replacing oxygen cylinders, and testing the R/T set would go on simultaneously. On many occasions all the aircraft of a squadron formation were replenished with fuel and ammunition and prepared for another battle within eight to ten minutes of landing. Regular repair and maintenance works were carried out day and night with all maintenance personnel pooled on each station. Most of the maintenance work on the signals equipment of the fighters had to be carried out at night by the light of torches. Trouble was caused by damp, and extreme care was necessary to keep R/T sets dry. Dispersal of aircraft not only to satellite aerodromes but over wide areas at each of them, combined with lack of transport, increased the labour requirements for a given job.

The blackout and damage to power and water mains and the station organisation added to the difficulties. Initially, bomb craters and building repairs were undertaken by station works and buildings detachments. As the damage grew, these units could no longer cope, and works repair depots were called in. Invariably a lot of the crater filling and rubble clearing was carried out by the airmen of the station itself. This was particularly important in the early days of the battle when the *Luftwaffe* concentrated its attacks on RAF installations, and sterling work was done by both RAF and RE teams.

The repair of damaged aircraft was another remarkable engineering achievement. Largely the responsibility of the Civilian Repair Organisation (CRO), 'miracles of repair were worked in 1940', with the weekly output of repaired aircraft reaching 160 planes a week. Between July and December 1940 the CRO put back 4,196 damaged planes; of the aircraft issued to fighter squadrons during the Battle of Britain, only 65% were new, with 35% being repaired aircraft. For these reasons, coupled with heroic work in the factories manufacturing the planes, British resources increased in comparison to the Germans', except for a short period towards the end of August. Indeed, the RAF finished the Battle of Britain with more fighters than when it started.

The Battle of the Beams

To enable their bombers to locate their targets the Germans used radio beams which intersected over the target and emanated from transmitters located along the coastline from Norway through the Low Countries to Brittany. Known as *Knickebein* (crooked

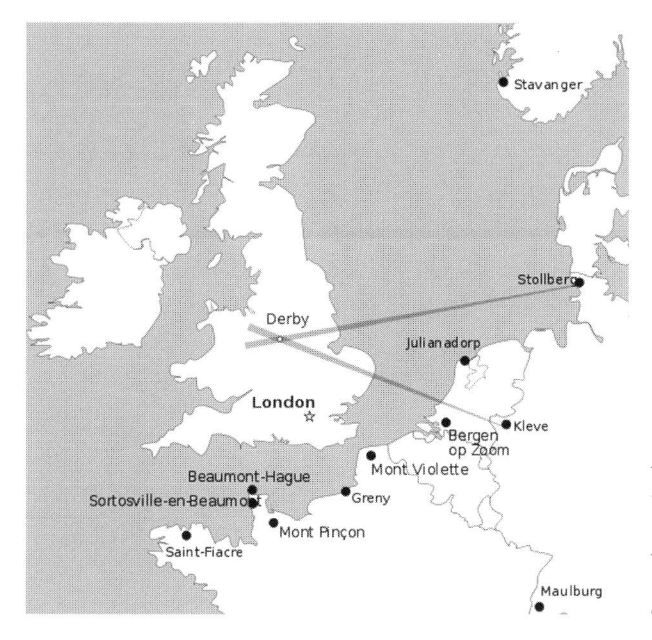

Early Knickebein transmissions. Transmitters were later located in Norway and France, thereby improving the system's accuracy. Wikipedia

leg), so named from the shape of the transmitting aerials, the beams were only a few metres wide. However, with the transmitters mounted on a rotating base, the beams could be transmitted in any desired direction. When a target had been chosen, two transmitters were set up so that their beams intersected over the target. Attacking aircraft would follow one of the beams until they picked up the signal from the other beam, when they knew that they could drop their bombs.

Air Ministry scientists led by Dr Reginald Jones became aware of the possibility that beams were being used by investigating the equipment on a bomber which had been shot down, by eavesdropping on the conversations of prisoner-of-war pilots, and by reference to 'bombing beams' in a Bletchley Park decrypt. To investigate whether this was a reality, an RAF bomber was fitted with appropriate radio receivers and told to search for radio signals. Two signals were detected, and when they were plotted they were found to intersect over the Rolls-Royce factory at Derby – at that time the only factory producing the Merlin engine which powered the British fighters.

Countermeasures were surprisingly easy and became more sophisticated with experience. RAF Anson aircraft fitted with appropriate receivers flew so as to pick up the German *Knickebein* signals. When they were discovered, a message was sent to a British radio station near the intended target which then emitted its own similar signals that the German aircraft also picked up. The German navigators were unable to tell which signal was which and they drifted off their intended course – in effect the British 'bent' the German beams. Eventually the British were able to 'deflect' the beams by a specific amount, thereby enabling them to intersect over open countryside. The

Jones, Dr Reginald V. CH CB CBE FRS (1911–97)

After reading natural sciences at Oxford, Jones joined the Royal Aircraft Establishment (RAE), Farnborough, in 1936 and in 1939 was transferred to the Air Ministry's Intelligence Section where he was involved in assessing German technology and developing countermeasures. His first job was to study 'new German weapons', real or potential, including the *Knickebein* radio navigation system which directed German bombers to their targets. Following investigations by Jones and the RAF the British were able to build jammers whose effect was to 'bend' the beams with the result that the bombs were dropped over open country. Thus began the famous 'Battle of the Beams' which lasted throughout much of the war, with the Germans developing new radio navigation systems and the British developing countermeasures to them. Jones frequently had to fight against entrenched interests in the armed forces, but in addition to enjoying Churchill's confidence, he had strong support from Frederick Lindemann and the Chief of the Air Staff Sir Charles Portal. As far back as 1937 Jones had suggested that metal foil falling through the air might create radar echoes which would appear on enemy radar screens as 'false bombers', a concept which evolved into 'window' or 'chaff'. Unbeknown to the British, the Germans also knew about it and both parties were reluctant to use it out of fear that their enemy would do the same; this delayed its deployment for almost two years. Jones also served as a V2 rocket expert on the Cabinet Defence Committee (Operations) and headed a German long range weapons targeting deception under the Double Cross System. He is regarded as the 'Father of Scientific and Technical Intelligence'. In 1946 he was appointed to a chair at the University of Aberdeen which he held until his retirement in 1981. Moving away from Intelligence his work at Aberdeen concentrated on improving the sensitivity of scientific instruments.

countermeasures also confused the German navigators so that some were uncertain how to return to their bases – a few actually landed at RAF airfields!

The Germans also developed more sophisticated radio navigation aids known as *X-Gerät* (X-Apparatus). This was used for the raid on Coventry in November 1940, and Jones devised countermeasures after analysing the equipment on a plane shot down over Dorset. They also employed *Y-Gerät*, which was also sometimes referred to as *Wotan*, the name of a one-eyed god in German folklore. As the Germans often used codenames which gave an indication of the process being referred to, Jones correctly inferred that this was a single directional beam, probably operating in conjunction with a return signal from the aircraft which gave a distance measurement. This enabled German ground control engineers to determine the aircraft's precise position and thus direct it to its target.

By chance, the Germans had chosen the operating frequency of the *Wotan* system to be the same as that of the powerful but dormant BBC television transmitter at Alexandra Palace. Jones then arranged for the return signal from the aircraft to be intercepted and sent to Alexandra Palace for re-transmission. The combination of the two signals made the system virtually inoperable. As *Wotan*'s use went on, the aircrew accused the ground station of sending bad signals and the ground station alleged the aircraft had loose connections. The Germans did not realise that the system was being interfered with, instead believing that it suffered several inherent defects; they eventually lost faith in electronic navigational aids and did not deploy them again against the UK.

Flying by Night

During the early months of the night raids, British responses were poor and largely ineffective. Among the reasons for this was that the radar network which had helped win the Battle of Britain was ineffective at night because it ultimately depended on visual sighting of the enemy by the fighter pilot. There were also too few searchlights and they were generally useless, particularly against high-flying aircraft or if there was low cloud cover. Although all the available anti-aircraft guns were busy firing there were also too few (some 1,500 compared to the 4,000 considered necessary), and the vast majority depended on visual sightings – again, difficult at night. But, as they gave the civilian population the impression that something was being done to deter the bombers, they helped maintain civilian morale. Lieutenant-General Pile wrote subsequently: 'Our London (gun) barrage was a policy of despair. Every gun had been told to fire every available round on an approximate bearing and elevation. Morale had been saved, if there had been no other result.'

Pre-war research into night defence had been concentrated on attempts to light up the cloud cover so that planes were silhouetted against it – and the funds allocated to that were 40 times greater than those initially allocated to radar research. Acoustic and infrared methods were also investigated.

One of the members of a team set up under Watson-Watt at Bawdsey Park in 1936 to investigate airborne radar possibilities was Dr Edward 'Taffy' Bowen. He was one of the two scientists who conveyed the hardware of the Tizard mission across the Atlantic (see Chapter 1).

Initially the team focussed on investigating the feasibility of detecting surface ships by airborne radar, and by June 1937 they had demonstrated that it was possible to detect a 2,000-ton freighter at a distance of 6 to 8 km. It was not until May 1939 that investigations began on Airborne Interception (AI) radar possibilities. Twelve months

Bowen, Edward 'Taffy' CBE FRS (1911–91)

After graduating from Swansea University in 1930 Bowen studied under Professor Appleton at King's College London where he gained a doctorate. In 1935 he was recruited by Watson-Watt to join his team at the Radio Research Station investigating the possibilities of radar. He was accidentally responsible for saving the research that culminated in the daytime radar chain, as during air exercises in 1936, his was the only receiver to recognise the signals reflected from an aircraft. All other receivers were nearer the ground and the signals were blanketed by trees. That incident apart, he concentrated on developing airborne radar – used with increasing success against German bombers during the Blitz – and in locating surfaced submarines. By late-1940, Rowe having replaced Watson-Watt as senior scientist, Bowen had been largely side-lined and went to the USA with Tizard to brief the Americans on radar developments. In 1943 he went to Australia to work at their radiophysics laboratory, and in 1946 he was appointed head of the relevant division. He also worked on the new science of radio astronomy and was instrumental in both obtaining American funding for the large radiotelescope built at Parkes, New South Wales, and in developing radio astronomy in the USA. He had an enduring love of cricket and played regularly. After suffering a stroke he died in 1991, aged 80.

later in May 1940 a Fighter Interception Unit (FIU) was established to develop airborne radar as a tool in attacking enemy aircraft. In addition to the need to create effective electronic equipment, it was vital that the transmitter, receiver, and screen could be made small and light enough to be mounted in a small aircraft (the former attached to the fuselage and the latter two inside alongside the operator).

The cavity magnetron and centimetric radar were the vital ingredients in making radar sets both small enough and capable of spotting other aircraft. Originally invented by American scientist Albert Hull in 1920, the cavity magnetron was little used until British scientists John Randall and Harry Boot of the University of Birmingham were able to modify and improve it in 1940. Described as the single most important invention in the history of radar, it was a small device that generated microwave frequencies, thus allowing the development of centimetric radar and of sets small enough to fit inside aircraft. It was one of the most important of the secrets taken to the United States by the Tizard Mission.

The first 'kill' had been achieved in November 1940 by a Blenheim aircraft, but as the technology improved the equipment was also installed in other aircraft, particularly the Bristol Beaufighter, the name given to the Beaufort night-fighter. It was only when

Cavity magnetron.
Science at War (HMSO)

GCI radar working with the AI radar sets, which went into service in March 1941, that German bomber losses started to increase.

Successful aerial night defence by radar was entirely due to the persistence of Bowen, who overcame major problems to create AI radar as well as ASV radar, both of which were world firsts.

Bad weather during January and February 1941 reduced the number of German raids and gave the British time in which to train both aircrews and ground crews – a necessity before the combination of Beaufighter, AI radar, and GCI radar could be integrated into an effective fighting unit. By March the RAF was prepared; and, as the following table shows, the number of aircraft shot down increased dramatically until the raids ceased in mid-May when the *Luftwaffe* began preparing for the invasion of Russia.

Table 1 *German night raiders destroyed, March–May 1941*

Month 1941	**Aircraft Destroyed**	**Aircraft probably destroyed**	**Aircraft damaged**	**Total aircraft casualties**
March	27	9	12	48
April	53	18	25	96
May	111	25	54	190
Totals	191	52	91	334

In his book, *One Story of Radar*, Albert Rowe wrote: 'The night war of the air was the second major British victory of the war and, as with the first, victory depended upon radar. This second victory was achieved with the aid of a few GCI sets and perhaps not more than 100 AI sets.'

Although it would have been feasible for night radar to have been introduced earlier, professional rivalries, misplaced belief in alternatives, and difficulties in perfecting the equipment prevented its effective use before March 1941. Bowen was awarded £12,000 (awards No. 178–187) by the Royal Commission on Awards to Inventors.

Pillboxes and other Obstacles

That the situation facing Britain in June 1940 was dire is evidenced by the fact that in the St Margaret's Bay region near Dover, a prime potential invasion area, there were only three anti-tank guns to defend some 7 km of coast and each gun only had six rounds of ammunition.

Following the Dunkirk evacuation, and while the Battle of Britain was raging overhead, the RE masterminded the erection of rudimentary anti-invasion defences on possible air and sea landing sites, both around the coast and inland. These included the construction of pillboxes and anti-tank obstacles, as well as other obstructions to thwart airborne and seaborne landings on beaches, estuaries, and potential parachutists' landing sites. Minefields were also laid. The speed at which they were laid, coupled with the inexperience in such duties of many of the troops involved, caused difficulties later when the fields had to be moved or cleared.

Typical coastal anti-invasion defences, one of several design types. These are to be found on the Isle of Grain and comprise flat-topped pyramids that were called dragon's teeth. Tony Watson. See also http://hoo-peninsula.blogspot.com/2011/02/isle-of-grain-ww2-anti-tank-obstacles.html

Three defence lines were built in arcs to defend the coast, London and the Home Counties, and the industrial heartlands of the Midlands. The anti-tank structures were mostly concrete blocks situated alongside roads, with sockets built into the roads into which steel stanchions up to 1.5 metres high could be placed if the road had to be closed following an actual invasion. Camouflaged pillboxes from which troops could fire on an advancing enemy were also built in strategic positions nearby. Many are still visible in various parts of southern and eastern England.

The RAF attacked invasion barges that were being assembled in Channel ports on the Continent and had destroyed over 200 by mid-September.

When it became apparent that the Battle of France was about to be lost, Britain established the Home Guard (originally called the Local Defence Volunteers and later nicknamed 'Dad's Army'), principally for veterans and youths to defend their neighbourhood in the event of an invasion.

Auxiliary Units were also created; comprising young farmers, gamekeepers and others with extensive knowledge of the locality, who would in the event of an invasion operate as resistance fighters and saboteurs working behind enemy lines. They had a life expectancy of about two weeks. The RE constructed underground bunkers in woodland areas as bases from which the units would operate. These were equipped with arms, explosives, food, and bunks – and with a camouflaged entrance and an emergency escape tunnel. Some additional observation posts also had a special and compact high frequency radio operating on a non-military waveband which would be used by male or female civilian or military personnel.

The RE were also responsible for masterminding the design and construction of the camps and other bases needed to accommodate Britain's expanding military establishment, as well as the troops from Canada and other Commonwealth countries who were arriving in the UK. Much of the detailed work was undertaken by civilian engineering consultants, architects, and contractors.

The vastly greater urgency and volume of signal traffic that had to be handled led to significant changes in organisation and technique. Throughout the Battle of Britain the Defence Telecommunications Network proved invaluable, providing teleprinter and other links between radar stations, Observer Corps lookouts, Fighter Command headquarters, and the operational airfields. During the anti-invasion phase of 1940 special communications systems were also put in place to link higher commands with the outer defence lines and the anti-ship radar network. Even a pigeon service, which had been used during the First World War, was hastily reactivated for a short period.

Surviving the Blitz

In September 1940 the Luftwaffe switched its attacks from RAF bases in south east England to London's East End, initially with over a thousand aircraft attacking day and night dropping both high explosive and incendiary devices. By the end of the month,

Bridge built at Bank Station by the Royal Engineers. London Metropolitan Archives, City of London

however, after sustaining unacceptable daytime losses, the *Luftwaffe* concentrated on night-time attacks.

By November other parts of London and other major cities, ports and factories were being attacked. Coventry's city centre was largely destroyed in a raid by 500 aircraft on the 14th, while London was attacked for 76 consecutive nights, the incendiary bomb attack of 29 December causing the 'Second Great Fire of London'. All major towns and ports suffered extensively, the second most heavily bombed city being Liverpool, which suffered a full week of intensive bombardment in May 1941.

Although some raids continued after May, the worst of the aerial bombardment was by then over. British countermeasures had become more effective and the Germans realised that their attacks had not achieved their aims of subduing the population into seeking peace and of seriously affecting war production.

Roads, bridges, and railways were destroyed or damaged though, and as the telephone, electricity, gas, water and sewerage networks were disrupted (many municipally owned and operated) it fell to the local civilian engineers to take the lead in repairing and reinstating them.

Such engineers included water engineers, whose work was vital in ensuring that the water network could provide water, not only to the population but also to the firefighting service in the quantities and pressures they needed. Also involved were Post Office, electricity, gas, and railway engineers who maintained and repaired telecommunication networks, energy supplies, and rail tracks and bridges, as well as engines and rolling stock. Moreover, municipal engineers, in addition to specific responsibilities for roads, bridges, housing, and so on, had overall coordinating responsibilities.

Typical of the latter was Thomas Frank, London County Council's chief engineer. He was coordinating officer for road repairs and public utility services for the London area, and he directed the repair services that enabled London to carry on in spite of the severest air raids – a task for which he was knighted in 1942. Conscious that Central

Frank, Sir Pierson FICE (1881–1951)

Frank trained as a municipal engineer in Yorkshire and was appointed city engineer of Ripon in 1908. After four years there, he moved in succession to Stockton-on-Tees, Plymouth, Cardiff, and Liverpool in each of which he served for about four years as the city engineer or equivalent. During the First World War he served with the Royal Engineers and was wounded in 1917. He was appointed city engineer and county surveyor to the London County Council (LCC) in 1930 and over the next nine years was responsible for many significant works in the capital, including the demolition and replacement of the old Waterloo Bridge in which he was assisted by consulting engineers. During the Second World War he was co-ordinating officer for Road Repairs and Public Utility Services for the London area, and directed the repair services that enabled London to carry on in spite of the severest air raids. He was knighted for this in 1942. Frank was very conscious that Central London and its underground transportation system could be vulnerable to flooding were its river walls to be breached; he established rapid response teams ready to undertake speedy repairs should this happen. He retired from the LCC in 1946 aged 65 and joined consulting engineers Coode, Vaughan-Lee and Gwyther, with whom he worked on a variety of projects at home and abroad. Among his many professional and other appointments were being colonel in the Engineer and Railway Staff Corps and being elected president of the ICE in 1945.

London and its underground transport system would be vulnerable to flooding were its river walls to be breached, Frank established rapid response teams ready to undertake speedy repairs, something which was top secret lest news of it might precipitate such an attack.

In 1941 a bomb destroyed Bank Underground Station in Central London and also created a massive crater at the junction of six important roads. Within 90 minutes of the incident, sappers and pioneers had started clearing the site prior to the construction of a large box girder bridge. Capable of carrying London Transport buses nose to tail, it was built in two spans of 15.3 and 34.4 metres and was completed in four and a half days.

External Anderson and internal Morrison shelters for placing in gardens or under tables were designed, and large numbers were distributed. Despite their simplicity, they were very effective in reducing casualties. In London, some tube stations and short lengths of line were also used as shelters. Persons who worked in important urban buildings were formed into teams of fire-watchers who slept on site and were provided with stirrup pumps and buckets of sand so as to disable incendiary devices.

When a town was attacked firefighting appliances from neighbouring towns would be rushed in, although initially they were often inoperative because there was no standardisation of hydrants and hoses – a situation that persisted until adaptors were made available.

Thanks to the Bletchley Park decrypting operations it was often known which town would be attacked on a particular night and, after the raid had started, small diversionary fires were deliberately started in the neighbouring countryside in the hope that bombs would fall there rather than on the town.

Frequently hampering operations were the unexploded bombs (UXB) that, with great bravery and loss of life, RE bomb disposal units working under Major-General G. B. O. Taylor defused. Their naval and air force counterparts did similar work at ports, estuaries, and airfields. Although the majority of bombs dropped exploded, a number did not, either because they were faulty or because they had been fitted with delayed action fuses. If this happened near homes, factories, or railway lines, the population had to be evacuated and production and transportation came to a halt.

By September 1940 some 10,000 RE personnel had been formed into 400 bomb disposal sections based in London, other major ports and cities, and elsewhere around the country. They wore a special cuff badge and, for speedy recognition, their vehicle mudguards were painted red. Often the bombs penetrated the ground to a considerable depth, and the team's first task was to excavate down to expose the bomb so the officer in charge could examine it and decide what action should be taken.

By the end of the war the number of unexploded items dealt with totalled 45,441 bombs, 6,983 anti-aircraft shells and nearly 300,000 beach mines. Total RE casualties were 55 officers and 339 other ranks killed, and 37 officers and 172 other ranks wounded. As their work was not 'in the face of the enemy' their bravery could not be recognised by the award of the Victoria Cross or other gallantry medal. To overcome this, the George Cross and George Medal were instituted. Thirteen officers were awarded the George Cross (two posthumously, with three dying later), and many more the George Medal.

Typical of many of the brave men who served in these units was Lieutenant (later Colonel) Stuart Archer RE, who was awarded the George Cross for tackling unexploded bombs in South Wales in the summer of 1940. In the early months of the *Blitz* there was little knowledge of how the fuse and booby-trap mechanisms worked, and if a bomb was successfully defused the salvaged parts were sent to the War Office for examination by their experts. Over the next three months Archer safely detonated over 200 bombs as well as successfully extricating five fuses and a new booby-trap mechanism that had claimed the lives of a number of other UXB personnel. He was

Archer, Colonel Stuart GC OBE FRIBA RE (1915–2015)

The son of an electrical engineer, he received his secondary education at the Regent Street Polytechnic School of Architecture. Having qualified as an ARIBA in 1936 he was employed by a firm in Gray's Inn in which he later became a partner. He joined the Honourable Artillery Company pre-war but in March 1940 was commissioned in the Royal Engineers and assigned to a bomb disposal unit. Three months later he was posted to South Wales where he successfully neutralised over 200 unexploded bombs and retrieved a number of fuses and booby-trap devices. He was awarded the George Cross in 1941. Having been demobilised in the rank of major he returned to his architectural practice but a few years later joined the Army Emergency Reserve with whom he served until its disbandment in 1967. He was appointed OBE in 1963, elected a fellow of the Royal Institute of British Architects in 1970, and served as chairman of the VC and GC Association from 1994 to 2006. He died in May 2015 aged 100.

awarded the George Cross for his bravery and in 1994 became the chairman of The Victoria Cross and George Cross Association. An architect by profession, he died aged 100 in 2015.

In many of these activities women played as vital a role as their male counterparts...

Chapter 5

WOMEN AT WAR

In the services, planning the national diet, advancing medicine and boosting production

Wars had traditionally been regarded as all-male affairs. However, this was not the case in the Second World War when women in each of the Allied countries made major contributions towards the victorious outcome of the war. In the UK from December 1941 they were conscripted to serve in the armed forces, industry, or other essential services. The female branches of the armed services were: the Women's Royal Naval Service (WRNS but almost always referred to as Wrens); the Auxiliary Territorial Service (ATS) and the Women's Auxiliary Air Force (WAAF). Large numbers served at Bletchley Park, for example, and during the *Blitz* many worked with distinction as ARP (Air Raid Precautions) wardens, as ambulance drivers, or in the National Fire Service. There were many others in hospitals and other medical organisations; in factories producing airplanes, weapons, munitions, and various machines; or on the land on farms or in forestry work. Many continued working as housewives and mothers, also contributing to the war effort in a variety of ways.

In addition to performing many conventional female tasks in catering and secretarial work, women replaced men in many key roles and were responsible for essential operations, such as working as plotters in Fighter Command's headquarters

A QO (Quick-Firing Ordnance) Wren removing a machine gun for maintenance. IWM

Anti-aircraft gunners in London take a break. IWM

during the Battle of Britain. Wrens crewed small boats in harbours while members of the ATS served in searchlight and anti-aircraft units. Among the women who joined the ATS was Princess Elizabeth, now Queen Elizabeth II, who served in a vehicle maintenance unit. Many aircraft pilots joined the Air Transport Auxiliary (ATA) and ferried unarmed aircraft from factories to RAF airfields. Towards the end of the war their members also ferried planes across the Atlantic. They also transported senior service personnel and some performed air ambulance work.

Lots of the operatives sent to France and other occupied countries by the Special Operations Executive (SOE) were women – a number of whom were captured, tortured, and executed – while most of the personnel working at Bletchley Park were women. Here they served both as cryptanalysts and as the Wrens who operated the decoding machines. A few Wren officers were also decrypting officers in the troopships converted from the fast peacetime passenger liners whose speed enabled them to cross the oceans unescorted. If Bletchley Park decrypts suggested U-boats might be approaching, signals were sent instructing them to change course.

Each service also had its own nursing organisation whose members served in every theatre of war, as did nurses of the First Aid Nursing Yeomanry (FANY). At home civilian female doctors, nurses, and other medical auxiliaries provided medical support

WAAF plotters and RAF controllers in Operations Room, Group 10 HQ Wiltshire. IWM

as required, along with members of the St John Ambulance Service. Many women also staffed the service canteens which were run by NAAFI (the Navy, Army and Air Force Institute).

Formed in 1938, the Women's Voluntary Service (WVS) already had 165,000 members by the start of the war. Originally intended to provide assistance in the event of air raids, its members' work quickly diversified into helping in all areas of the Home Front. By the end of 1941, the WVS had enrolled its millionth member. They provided support, food, rest centres, washing facilities, and new clothes for bomb victims; staffed Incident Inquiry Points to give information about the dead and injured to relatives and friends; ran mobile canteens for firemen and ARP rescue workers; staffed hostels, clubs, and communal feeding centres called 'British Restaurants'; and undertook welfare work for troops. Volunteers organised salvage drives to generate raw materials for the war effort, including the collection of aluminium saucepans and kitchen utensils and the removal of gates and iron railings from private and public buildings. They also coordinated clothing exchanges where children's clothes could be swapped for larger sizes. They ran some 200 British Welcome Clubs to bridge the divide between British civilians and the large numbers of American troops arriving in the country prior to the invasion of France.

Among the other women's organisations were the Women's Land Army (WLA) and the Women's Timber Corps (WTC) whose members were commonly known as 'Lumber Jills'.

In the USA, nearly 350,000 women served in uniform both at home and abroad. General Eisenhower acknowledged that 'the contribution of the women of America, whether on the farm or in the factory or in uniform, to D-Day was a sine qua non of

Platt, Beryl Catherine, Baroness Platt of Writtle (née Myatt) CBE DL FRSA FREng HonFIMechE (1923–2015)

Raised in Southend-on-Sea, after schooling there and in Slough she went to Cambridge in 1941, where she was one of only five young women studying engineering (compared to 250 men). Leaving Cambridge in 1943 she joined the Hawker Aviation Company; whilst there she worked for Sydney Camm on the production and testing of the company's fighter aircraft. Post-war she continued with the company focussing on safety issues for passenger airliners until her marriage in 1949, which saw the end of her professional engineering career. Shortly afterwards she became active in local politics and was elected a Conservative member of Essex County Council and served as its chairman in the 1970s. In 1981 she was elevated to the peerage and in 1983 was appointed chair of the Equal Opportunities Commission. She was instrumental in establishing WISE (Women into Science and Engineering), an organisation created to highlight the career opportunities for girls and women in the scientific and engineering professions. She was the first female member of the Engineering Council and only the second woman to be elected to the Fellowship of Engineering (now Royal Academy of Engineering). She was an active member of the House of Lords and served on the Select Committee on Science and Technology.

the invasion effort.' Russian women also had a significant impact both as civilians and in the armed forces, where they were more actively involved than in other combatant countries.

Others with technical qualifications worked in research establishments and design offices. One such was Beatrice Shilling, whose invention of a modification to the Merlin engine's carburettor made a major contribution towards winning the Battle of Britain (see Chapter 4). Another was Beryl Myatt (later Baroness Platt of Writtle), who, after reading engineering at Cambridge, joined Hawker Aircraft Limited in 1943 where she worked on the testing and production of the Hawker Hurricane, Typhoon, and Tempest fighter planes. Post-war she entered politics and did much to advance the role of women in technology.

When children were evacuated from the major cities at the start of the war, they were accompanied by their teachers, many of whom were women. Many of the artistes who, as members of ENSA (Entertainments National Service Association), entertained the troops were women. These included Gracie Fields and Vera Lynn, the 'Forces' Sweetheart', whose songs included 'The White Cliffs of Dover,' 'We'll Meet Again,' and 'Lili Marleen,' the German love song much liked by British as well as German troops.

As in other walks of life, alongside most of the political and military leaders was a woman whose input into her husband's career and successes was generally invisible but nevertheless very real. A reviewer of Sonia Purnell's recently published biography of Clementine Churchill said, 'Churchill would never have risen to greatness without Clementine.'

Feeding the Nation

Housewives, together with women working in civilian tasks, had to cope with the effects of food and clothes rationing, the former in particular requiring great ingenuity in eking out the food that was available.

Nutritionists Elsie Widdowson CH CBE FRS (1906–2000) and Robert McCance FRS (1898–1993) formulated the wartime national diet which was designed to optimise public health while ensuring that food imports were minimised and comprised only essential foodstuffs. Following the outbreak of war they and a number of colleagues became their own experimental subjects, eating a starvation diet of bread, cabbage, and potatoes for several months to find out if wartime rationing – with little meat, dairy, or calcium intake – would affect their health. They showed that good health could be supported by a very restricted diet, albeit with calcium supplements, and their work became the basis of the Ministry of Food's wartime austerity diet. Their work resulted in the first mandated addition of vitamins and minerals to food when calcium was added to bread.

Rationing improved the health of British people, infant mortality declined, and, discounting deaths caused by hostilities, life expectancy rose because everyone had

Food Production

Pre-war some two-thirds of Britain's food consumption was imported, with much beef coming from Argentina, lamb and butter from New Zealand, and grain from the United States.

The shipping losses caused by Germany's U-boat campaign necessitated:

- The introduction of food (and petrol) rationing
- Urgent measures to increase home food production; the 'Dig for Victory' campaign resulted in many parks, golf courses, allotments, and private gardens being used for vegetable production
- The manufacture of large numbers of tractors to increase farm production
- Research into establishing the optimum diet so as to ensure that the most nutritional foods were grown and imported

Widdowson, Elsie CH CBE FRS (1906–2000)

Widdowson grew up in south London and studied chemistry at Imperial College where she was one of the first female graduates. After obtaining a PhD for a study on the carbohydrate content of apples she went on to work at Middlesex Hospital on the metabolism of kidneys before moving to King's College London to study for a postgraduate diploma. In 1933 she met Dr Robert McCance, then a junior doctor researching the treatment of diabetes. They realised their research was complementary and they became collaborators on dietary issues over the next 60 years. In 1938 they both worked at the Department of Experimental Medicine in Cambridge where they formulated the wartime national diet. Working for the Medical Research Council in 1946 they studied the effects of malnutrition on civilians in Nazi-occupied countries and the inmates of concentration camps. Widdowson's later work included studying infant diets and malnutrition in Africa. She served as president of the Nutrition Society, the Neonatal Society, and of the British Nutrition Foundation and was elected a Fellow of Imperial College and of the Royal Society. She was awarded a CBE in 1979 and made a member of the Order of the Companions of Honour in 1993.

access to a varied diet with enough vitamins. Among a number of other wartime scientific investigations was one into the practicality of harvesting plankton from Scottish sea lochs to augment the food supply.

Medical Advances

A significant number of the doctors, nurses, and technologists involved in the provision of medical services were women, and several were closely involved in their various improvements. Medical research in the UK was coordinated by the Medical Research Council, which had been established in 1920, while the comparable body in the USA was the Committee on Medical Research created by Vannevar Bush in 1941. These two bodies collaborated closely during the war.

Penicillin was one of the many wartime medical developments that continued into the post-war era. Originally discovered accidentally by Alexander (later Sir Alexander) Fleming FRS (1881–1995) in 1928, it was thought to be too unstable for mass production as a drug. In 1940–41, however, an Australian physiologist, Professor Howard Florey, and Ernst Chain, a refugee from the Nazis working at Oxford, established its remarkable healing properties. Florey was authorised to go to USA to describe his discoveries and encourage mass production of the drug by the Americans. Meanwhile Imperial Chemical Industries (ICI) installed a manufacturing plant in the UK which produced supplies that were sent to the British Army in North Africa and yielded remarkable results. Manufacturing was then stepped up in both the UK and US and by the time of the Normandy landings there were sufficient supplies to treat all the casualties needing it.

Reviewing the achievements of the 14th Army in the Southeast Asian campaign, Field-Marshal Lord Slim said: 'The scientific ideas which helped us most were on the medical side. ... By the beginning of 1943 my army were steadily disappearing from disease, mainly malaria, but with a melancholy list of other diseases like scrub typhus, skin diseases and the rest.'

In the Macedonia campaign of 1916, 40% of the British force had gone down with malaria. Remembering this, Sir Edward Mellanby, the chief medical advisor to the War Cabinet, called for increased production of anti-malarial drugs as early as 1940, when British troops were fighting in North Africa. The situation was exacerbated early in 1942 by the Japanese occupation of the Dutch East Indies, the main source of quinine; which prompted an increased production of synthetic drugs, particularly mepacrine. A special manufacturing plant was developed, and although it involved a long and complicated process, British output rose from a mere 10 kg in 1939 to some 50,000 kg in 1943. After daily doses were made mandatory for troops operating in malarious zones, with severe penalties for defaulters, the numbers of hospital admissions decreased dramatically.

Bacillary dysentery had seriously affected the efficiency of troops fighting in the Middle East in the First World War; similar epidemics did not occur in the Second World War thanks to better sanitation, the reduction of flies, and the replacement of horses by mechanised transport. (This was unlike the German Afrika Korps, whose members suffered because of inadequate sanitary arrangements at their camps.)

Treatment was revolutionised by Anglo-American research and by the American manufacture of a new drug called sulphaguanidine. In New Guinea both Australian and Japanese forces were ravaged by an outbreak of the disease. All available supplies of the drug were rushed there and the epidemic was brought under control within ten days. The Japanese failed in their attempt to take Port Moresby and it was subsequently said that 'sulphaguanidine saved Moresby'.

Sir Archibald McIndoe CBE (1900–60) was a pioneering New Zealand plastic surgeon who greatly improved the treatment and rehabilitation of badly burned aircrew while working at the Queen Victoria Hospital in East Grinstead, West Sussex. Observing that aircrew that had landed in the sea healed more quickly than those who had come down on land, McIndoe developed the present-day practice of saline water treatment of burns. McIndoe was instrumental in forming The Guinea Pig Club in 1941 as a social and drinking club for those of his aircrew patients who had gone through at least two surgical procedures, together with the surgeons and anaesthetists who had treated them. By the end of the war the club had 649 members.

Heart surgery was one of a number of medical procedures which improved dramatically as a result of work undertaken during the war by surgeons serving with the Royal Army Medical Corps and the US Army's Medical Corps – and which has led to today's life-saving operations.

When the war started, the Germans and Americans had the most advanced masks for providing oxygen to air crews operating at heights of 9,000 metres and above. A team at the Harvard School of Public Health improved the American design and developed a mask which incorporated a microphone and withstood freezing and inward leaking of oxygen-free air at high altitudes. Shortly before the end of the war pressurised cabins were introduced into the B-29 Superfortresses and photo-reconnaissance Spitfires.

Making the Tools

Many tens of thousands of women worked in factories producing war materials and armaments; they are commemorated in Sheffield by the recently erected statue *Women of Steel*. See Appendix 6 for information on the mobilisation of British industry.

'Rosie the Riveter' was a cultural icon of the United States, representing the American women who worked in factories and shipyards during the Second World War, and she was commonly used as a symbol of feminism and women's economic power. Veronica Foster, (1922–2000), popularly known as 'Ronnie, the Bren Gun Girl', was a Canadian icon representing nearly one million Canadian women who worked in the manufacturing plants that produced munitions and material for the Second World War. The women of other Commonwealth countries also participated in similar roles.

On 9 February 1941 Churchill concluded one of his periodic broadcasts to the British people – one which was also aimed indirectly at the American nation – with these words:

Women of Steel statue in Sheffield. Stephen Canning

The other day, President Roosevelt (sent a letter) to me, and in it he wrote out a verse … from Longfellow, which he said, "applies to you people as it does to us." Here is the verse:

… Sail on, O Ship of State!
Sail on, O Union, strong and great!
Humanity with all its fears,
With all the hopes of future years,
Is hanging breathless on thy fate!

What is the answer that I shall give, in your name? Here is the answer which I will give to President Roosevelt: Put your confidence in us. Give us your faith and your blessing, and, under Providence, all will be well. We shall not fail or falter; we shall not weaken or tire. Neither the sudden shock of battle, nor the long-drawn trials of vigilance and exertion will wear us down. Give us the tools, and we will finish the job.

The following month the United States enacted the Lend-Lease Act, and for the rest of the war American industry was the main arsenal and manufacturer of the equipment and weapons used by the Western Allies.

Harnessing American Industry

At the start of the war all British-owned foreign assets were taken over by the government and used to purchase armaments, food, and other supplies from foreign countries, in particular the United States. By early 1941 the reserve of assets was running dangerously low and in March that year President Roosevelt signed the Lend-Lease Act, whose formal title was 'An Act to Further Promote the Defense of the United States'. Under the Act the United States supplied Britain and other Allied nations with food, oil and armaments in exchange for leases on British Overseas Territories that they could use during the war. Although most of the goods were supplied free some (e.g. certain ships) had to be paid for or were leased and were returnable to the US after the end of the war.

The sudden, and unexpected, termination of the Lend-Lease Act in August 1945 caused major problems for Britain. With around 55% of her GDP being directed towards military activities, her ability to export goods whose sale could be used to purchase the essential products which had to be imported was seriously impaired. As a result the Anglo-American Loan Agreement was signed in August 1946, under which dollars were made available to the UK that were used to pay for essential imports in subsequent years as well as the lend-lease items in transit to the UK or which were already in Britain and for which payment was due under the old Act.

It was agreed that the repayment of the loan should be made over a 60-year period. A similar arrangement was also made with Canada. (These events and the background to them were largely the reasons for Britain's post-war foreign-exchange problems.)

Britain's continued survival therefore depended on maintaining her Atlantic lifeline. Operating from their new bases in western France, however, German U-boats posed an ever increasing threat …

ATLANTIC TIMELINE

1939

Sep	SS *Athenia* and HMSs *Courageous* and *Royal Oak* hit by torpedoes and sink.
Nov	HMS *Rawalpindi* (armed merchant cruiser) sunk by battlecruisers *Scharnhorst* and *Gneisnau* but their proposed foray into the Atlantic abandoned.
Dec	Battle of the River Plate; *Graf Spee* scuttled.

1940

Jan→	Magnetic mines neutralised.
Jul→	U-boat HQ moves to Lorient; U-boat bases built at French Atlantic coast ports. RN's Western Approaches HQ moves to Liverpool. UK acquires 50 old US destroyers.

1941

May	*Bismarck* sunk.
July	Germany invades Russia. Arctic convoys instituted.
Sep	US starts escorting convoys as far east as Icelandic longitude.
Dec	Germany declares war on USA. USN joins RN and RCN in Battle of the Atlantic.

1942–43

U-boats make many sinkings off US Atlantic seaboard and in Caribbean until convoys introduced.

Saint-Nazaire raid immobilises only dry dock on Atlantic seaboard which could service a German capital vessel.

US troops successfully escorted to North Africa for landings in Maghreb

Bletchley Park reads increasing numbers of German naval signals.

Application of Operational Research findings improves success rate.

New escort vessels enter service.

Escorts fitted with improved sonar, Hedgehog depth-charge launchers, AS radar, and voice-radio equipment.

Escort carriers, aircraft-carrying merchant ships, and long range aircraft fitted with ASV radar enter service and close mid-Atlantic air gap.

1943

May	U-boats withdrawn. Battle of Atlantic effectively over.

1944

Jun	Allied armies successfully conveyed to Normandy.
Jun→	RN and RAF continue to attack *Tirpitz* in her Norwegian bases.

1945

May	*Kriegsmarine*'s U-boats and other vessels formally surrender.

Chapter 6

ATLANTIC AGONIES

The U-boat menace – and its eventual defeat

For the redoubtable Winston Churchill, 'the only thing that ever really frightened me during the war was the U-boat peril.' He also claimed that 'the Battle of the Atlantic was the dominating factor all through the war. Never for one moment could we forget that everything happening elsewhere, on land, at sea or in the air, depended ultimately on its outcome.'

Lasting throughout the entire war, it is one of the longest battles in all history. Germany's main objective was to use her submarine fleet, and on occasion her major surface vessels, to attack the Allied convoys and lone ships sailing on the high seas, particularly from North America to the UK. They aimed to sever the maritime lifeline through which the food, fuel, armaments, and other vital imports on which Britain's very existence and survival depended were transported.

The U-boats frequently operated in wolf packs of up to 20 vessels; as the war progressed they became increasingly sophisticated, capable not only of crossing the Atlantic but also of being replenished at sea. Although the RAF's Coastal Command aircraft later sank a large number of U-boats, the Navy initially received little support from the RAF. From 1940 to mid-1943 shipping losses were heavy, particularly in the mid-Atlantic 'air gap' – the area which at that stage could not be reached by land-based aircraft or protected by carrier-borne planes.

Only two ships (HMSs *Rodney* and *Sheffield*) had been fitted with radar by September 1939, and it was late 1940 before the Home Fleet was fully equipped. (The aircraft carrier HMS *Ark Royal* did not have radar when it was sunk in the Mediterranean in November 1941.) By mid-1940 only some 40% of escort vessels had radar, and it was March 1941 before the more efficient centimetric radar sets started to be installed. At the start of the war no ship had High Frequency Direction Finding equipment. Radio telephony for naval use did not exist in 1939; it was July 1941 before the first set was installed in a ship.

U-boat Design

U-boat technological sophistication and capabilities improved consistently throughout the six-year campaign waged against Allied shipping, as did the strategy and tactics their commanders adopted. Construction of the new U-boat fleet began in 1935, and

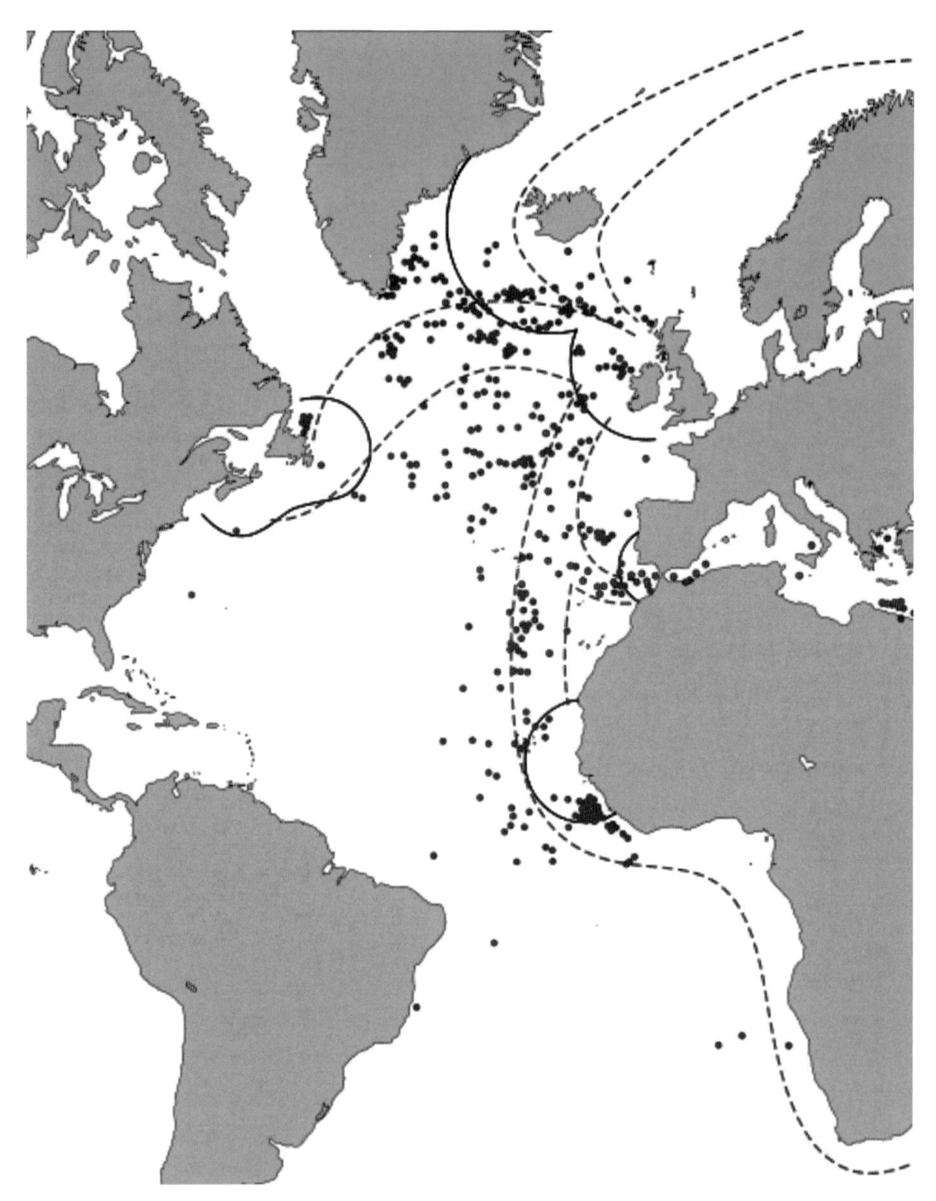

The pre-1943 Mid-Atlantic Air Gap. Air cover was only possible within the solid lines. Main convoy routes shown within broken lines. Sinkings shown as dots. Roskill: War at Sea.

by the start of the war the Germans had built 56 U-boats, of which 35 were operational. By the end of the war that had risen to 1,133, of which 782 were sunk by the Allied navies and air forces.

Apart from the earliest vessels, all U-boats were equipped with battery-driven electric motors, used to propel the vessel when submerged, and diesel engines, which

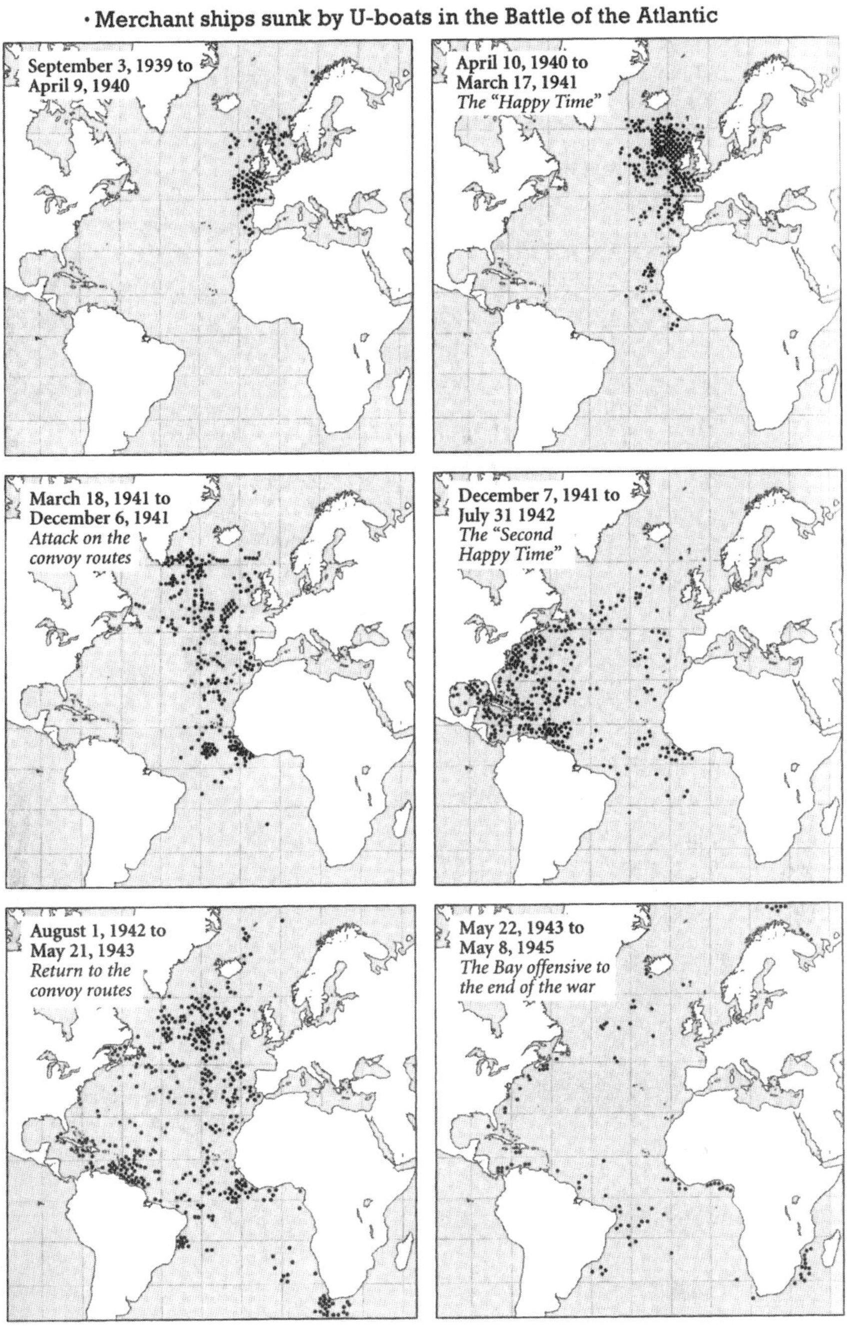

• Merchant ships sunk by U-boats in the Battle of the Atlantic

September 3, 1939 to
April 9, 1940

April 10, 1940 to
March 17, 1941
The "Happy Time"

March 18, 1941 to
December 6, 1941
Attack on the convoy routes

December 7, 1941 to
July 31 1942
The "Second Happy Time"

August 1, 1942 to
May 21, 1943
Return to the convoy routes

May 22, 1943 to
May 8, 1945
The Bay offensive to the end of the war

The Six Phases of the Battle of the Atlantic and locations of U-Boat sinkings of merchant ships. Maps by Dave Merrill, reproduced with permission from Stephen Budiansky, Blackett's War (New York: Knopf, 2013)

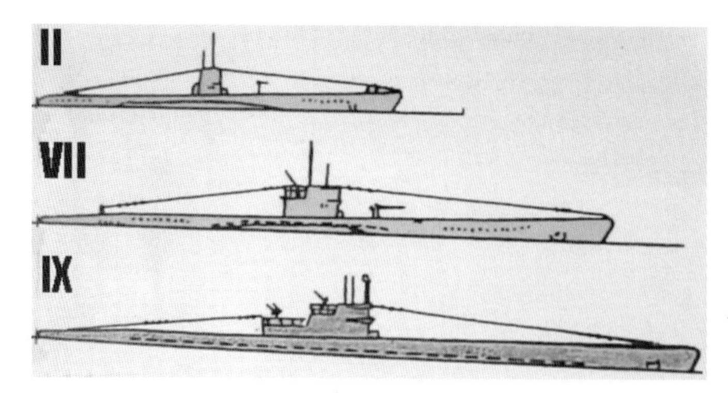

Type II – Coastal vessel Type IX – Long-range operations Type VII – The Atlantic workhorse. Anti-Submarine Warfare by David Owen

served the dual purpose of driving the vessel on the surface and recharging the electric batteries. Typical U-boats were as follows:

- Type II were coastal vessels constructed pre-war. They were 40 metres long, had a surface range of 4900 km and could carry five torpedoes or 12 mines. They had petrol engines.
- Most Second World War U-boats were Type VII. They were 67 metres long, carried 14 torpedoes, and had a crew of about 50 men. Their range and maximum speed were 13,600 km and 17.7 knots on the surface and 130 km and 7.6 knots submerged.
- Type IX were long-range vessels designed for operations on the American coast and in the South Atlantic and Indian Oceans. They were 76.5 metres long, could carry 22 torpedoes, and also had a crew of about 50 men. Their ranges and maximum speeds were 19,000 km and 18.2 knots on the surface and 100 km and 7.3 knots submerged.
- Type XXI U-boats had hydro-dynamically streamlined hulls, triple the number of battery cells, were quieter and could travel submerged for two or three days without recharging. They might have revived the U-boat campaign but for the fact that only two ever went into service.

Since higher speeds were possible, and because their diesel engines could obtain air and vent their exhaust fumes when on the surface, U-boats traditionally travelled on the surface by day as well as by night, only submerging when being attacked, likely to be observed, firing their torpedoes, or giving the crews respite in rough weather. Towards the end of the war the Germans developed the *Schnorchel* which, by providing tubes to connect the diesel engine's air intake and exhaust system to the atmosphere, enabled a U-boat to run its diesel engine underwater. It then needed to surface for only about four hours out of the 24.

While considerably increasing a vessel's range and operational capability, this system was unpopular with crews because the valve at the top of the air intake tube could close automatically in rough weather and leave them struggling for breath. Also, searching for targets had to be done through the periscope –which was difficult if the view was obscured by exhaust fumes. A number of Types VII and IX U-boats were equipped with *Schnorchels* and operated for many months. A potential development was a hydrogen peroxide-fuelled engine which would have significantly increased a U-boat's range and underwater speed, though it was introduced too late to be of value.

U-boat tankers, colloquially known as 'milk cows', were developed to refuel operational boats at sea, thereby significantly increasing their range and duration of operations. Allied escort vessels were instructed to give priority to attacking these vessels.

Instructions to U-boats were transmitted by radio from their headquarters. Up-to-date knowledge of the enemy's situation and tactics was a key component in the formulation of strategy and tactics and, in the same way that the Allied navies relied heavily on Bletchley Park decrypts, in the early years of the war the German decrypting organisation *B-Dienst* was often able to intercept and break both British and American naval signals. Following the development of ASV radar by the Allies, the Germans developed *Metox*, *Naxos* and *Hagenuk*, systems which could warn a U-boat of the approach of a radar transmitting aircraft by converting the radar signals into audible beeps.

As a counter to low-flying aircraft, some U-boats were fitted with anti-aircraft guns; since these were located behind the conning tower, attacking planes soon learned where they could not be shot at and their greater mobility enabled them to attack the gun crews.

Tackling the U-boat Menace

As in other battles, Allied success depended on a combination of strategy, tactics, luck, fighting spirit, and technology – and it was often a race to get a new idea from concept to effective application (see Appendix 2 for a description of the many stages involved.)

Much of the research undertaken by the Admiralty's various technical departments went into improving the means of detecting and attacking U-boats and of defending surface ships, both merchantmen and naval, against attacks from submarines and aircraft. Concurrently, a lot of effort also went into means of developing aerial attacks on U-boats.

When Bletchley Park broke the German Enigma code in early 1941 the Admiralty could pinpoint the locations of U-boats and was able to re-route convoys so as to avoid them. As new code books were issued every three months this initial success did not last long, but it was later repeated on a more permanent basis.

The German naval Enigma signals were more complex than those for the other two armed services, and their regular and speedy decryption involved the development of the world's first computer by Alan Turing and Thomas Flowers (see Chapter 2).

Before this became available, decryption had been achieved by brave naval personnel boarding a sinking U-boat to acquire vital equipment and documents. However, with the German operational system changing periodically, some of the information gained had a limited duration.

In consequence naval signals were read only intermittently in 1941, though more consistently from December 1942 onwards. When they could be read, convoys were routed away from the known positions of wolf packs and other U-boats, resulting in dramatic reductions in the number of sinkings.

The escort vessels' main tool for locating U-boats was ASDIC, a name probably derived from the initials of the Admiralty's Anti-Submarine Detection Investigation Committee and later known as **sonar** (an acronym for **SO**und **N**avigation **A**nd **R**anging). Spotting by Allied aircraft and Bletchley Park decrypts also contributed. The Allies' principal anti-submarine weapons were depth charges – metal canisters containing high explosives set to detonate at a predetermined depth.

Originally developed during the First World War, sonar was improved by Admiralty scientists during the interwar years and considerably more so throughout the Second World War. It comprised a transmitter on the search vessel's keel which generated sound pulses or 'pings' that were reflected by a U-boat (or any other submerged object) and then identified by a receiver on board the search vessel. The U-boat's direction was indicated on a cathode ray screen on the search vessel, while its distance was calculated by the time taken for the sound to return to the ship.

Hedgehogs and Squids

Once located, the U-boat was attacked by depth charges. In the war's early years these were dropped off the hunter's stern when passing over the vessel's presumed position. However, because the engine noise would have been heard, in the time taken for it to travel to the U-boat's presumed location the U-boat was able to dive or change direction – probably both. For this reason U-boat 'kills' were initially relatively low.

But the situation improved dramatically in 1942–43 when Hedgehog depth charge launchers started to be fitted to the hunters. These consisted of an array of smaller depth charges mounted on spigot mortars which were fired ahead of the hunting vessel at the U-boat's presumed position. Developed by the DMWD, the Hedgehog was an adaptation of the spigot mortar conceived by the special research department MD1. Its successful use was dogged by problems caused by its overhasty introduction; it was frequently badly fitted or inadequately serviced, problems which were compounded by the lack of an instruction manual because of the extreme secrecy which surrounded the project.

Improved variants such as the Squid were developed later. This was ordered directly from the drawing board in 1942, rushed into service in May 1943, and installed on 70 frigates and corvettes. It usually comprised two three-barrel mortar units which

Reloading depth charges on a Hedgehog on board HMCS North Bay. *Library and Archives Canada Photograph MIKAN no. 3394476.*

automatically fired depth charges at a moment determined by a sonic range recorder. A clockwork time fuse was used to regulate the detonation depth. The salvos were set to explode some 250 metres ahead of the ship and above and below the target, the resulting pressure waves crushing the submarine's hull. If caught on the surface, a U-boat could also be rammed or fired on by a ship's gun.

Attacks from the Air

By 1943 a number of Escort (or '*Woolworth*') Carriers had become operational. These were shorter and slower than conventional carriers and were designed to provide air protection to convoys. A considerable part of their effectiveness was due to their aircraft forcing the U-boats to stay submerged to avoid detection.

In addition, a number of merchant ships, mostly tankers and bulk grain carriers, were converted to carry aircraft. Called MACs (Merchant Aircraft Carriers), they also carried cargos and remained under MN command, although with RN personnel on

HMS Audacity – *Escort carrier. IWM*

MV Rapana *a MAC converted from an oil tanker. IWM*

A Hurricane fighter on the catapult on board a CAM ship. IWM

board for aircraft operations and maintenance. A number of merchant ships, known as CAM ships (Catapult Aircraft Merchant) were also adapted to carry a rocket-propelled catapult for launching a single fighter which could engage enemy aircraft. (After an action the pilot had to abandon his aircraft and parachute into the sea!). Five vessels were also converted to become FCSs (Fighter Catapult Ships) which carried a number of fighter aircraft that were also launched by catapult. Their success was, however, limited.

By 1943 Coastal Command had acquired new long-range Liberators and other aircraft, ensuring the convoys in the mid-Atlantic area could be protected. These were fitted with highly sensitive ASV radar and the Leigh Light, a powerful carbon arc searchlight fitted on aircraft (it was the idea of Wing Commander Humphrey Leigh, an RAF personnel officer, following discussions with returning aircrews). These airplanes could detect the *Schnorchel* tubes and periscopes of submerged U-boats and thus were able to bomb them accurately by night as well as by day. This resulted in a dramatic increase in the number of U-boat 'kills'.

Meanwhile, the Americans developed the Airborne Acoustic Homing Torpedo, deceptively also known as the Mark 24 Mine or FIDO to conceal its true identity. Propeller driven with an electric motor, it contained four hydrophones which detected the noise of the target and guided the steering system. It was claimed that it sank 31 U-boats and damaged 15 more.

Operational Research (OR) was now also being used to improve the effectiveness of the Allied attacks on U-boats. Although this new science had initially been developed to support RAF operations, it proved to be equally effective in support of the war against U-boats. It was pioneered by Patrick Blackett and his principal assistant Evan Williams, who had previously worked together at the Royal Aircraft Establishment on a system for the magnetic detection of submarines.

Leigh light. IWM

Operational Research (OR)

This was a new branch of science developed during the war in order to improve the effectiveness of military operations. All three armed services established an Operational Research centre whose function was 'the collation and critical analysis of facts and quantitative data on military problems'. Each was headed by a senior scientist:

- Professor Patrick Blackett, Director of Naval Operational Research at the Admiralty
- Sir Charles Darwin, later Professor C. D. Ellis, Scientific Advisor to the Army Council
- Professor Sir George Thomson, Scientific Advisor to the Air Ministry

Each of these scientists sat on the Joint Technical Warfare Committee, a subcommittee of the War Cabinet whose remit was to provide senior officers with 'definite pieces of data on particular aspects of operations'.

Patrick Blackett was regarded as 'the Father of Operational Research', and he not only defined its general principles but was also instrumental in introducing it to all three forces. He was also able to recruit leading scientists to the OR teams; at Coastal Command for instance, they included no fewer than six FRSs, two of whom were future Nobel Prize winners.

Air Marshal Sir John Slessor, who became commander in chief of Coastal Command in 1943, wrote in the Foreword to C. H. Waddington's *Operational Research in WW2 – Operational Research against the U-boat:* 'A few weeks [as C-in-C] were enough to convince me of the truth that all senior Commanders of any Service should ... acknowledge ... that an Operational Research Section is an absolutely indispensable component ... of any big Command.'

The first application of this new science had been in connection with the development of radar, when it was appreciated that the system's effectiveness could be increased by speeding up communications between the radar operators and Fighter Command. A Bawdsey Park communications engineering scientist, G. A. Roberts, was deputed to investigate how the interval could be minimised. Concurrently another scientist, Dr E. C. Williams, undertook a comprehensive analysis of the performance of the various early-warning stations, and by comparing the best with the worst, was able to recommend how to improve the operators' technique. These activities led to the establishment of a specialised group of scientists within the Fighter Command control centre who were also charged with investigating an ever-widening range of subjects.

As a result of the group's effectiveness, similar sections were established, first at other RAF commands and then at Navy and Army commands. During the *Blitz* the Director of Research and Experiments at the Ministry of Home Security established a similar section to investigate and study the effects of the bombing. Mainly recruiting staff from the Cement and Concrete Association, it eventually comprised 120 observers in the field and 40 analysts. Frequently, OR was able to demonstrate that it was more effective – and cheaper – to modify the design and/or method of application of existing weapons rather than develop new ones, an idea championed by Blackett in particular. Some of OR's many successes recounted elsewhere in this book have demonstrated this principle.

Blackett, Lord OM CH FRS (1897–1974)

An American book about Blackett's wartime achievements had on its back cover, 'In March 1941 after a year of devastating U-boat attacks, the British War Cabinet called upon intensely private, bohemian physicist named Patrick Blackett to turn the tide of the naval campaign. Though he is little remembered today Blackett did as much as anyone to defeat Nazi Germany, by revolutionising the Allied antisubmarine effort through the disciplined, systematic implementation of simple mathematics and probability theory.'

Patrick Blackett entered the Royal Naval College, Osborne, Isle of Wight, in 1910 and on the outbreak of the First World War joined HMS *Carnarvon* as a midshipman. He saw action in the Battle of the Falkland Islands and later, when aboard HMS *Barham*, at the Battle of Jutland. In 1919 he was one of a number of young naval officers sent to the University of Cambridge to broaden their education. However, enthused by the scientific opportunities that academic life offered, he resigned his commission and became an undergraduate. After graduating in physics in 1921 he undertook academic research at Cambridge and then at the universities of London and Manchester.

In 1935 he was invited to join Tizard's Committee for the Scientific Study of Air Defence which oversaw the development of radar. In the pre-war years he played a leading role in aiding refugee scientists and was a powerful influence on other left-leaning scientists in encouraging them to support the military effort. It is very probable that Blackett was closely involved with the publication of the Penguin Special mentioned in Chapter 1. Following the start of the Second World War he was first appointed scientific advisor to Anti-Aircraft Command (where he applied Operational Research principles) before becoming head of RAF Coastal Command's Operational Research section. In late 1941 he moved to the Admiralty, later becoming its head of Operational Research.

After the war he moved to Imperial College where he continued with his research activities on cloud chambers, cosmic rays, and palaeomagnetism, which helped to provide strong evidence for continental drift. The crater Blackett on the Moon is named after him. He was awarded the Nobel Prize for Physics in 1948 and was elected president of the Royal Society in 1965. A lifelong socialist, he advised Harold Wilson's administration on technical matters and in the late 1960s was created a Companion of Honour, awarded the Order of Merit, and elevated to the peerage.

U-boat illuminated by aircraft's Leigh Light. 'Hunting the Hunters', Mark Postlethwaite, www.posart.com

Williams, Dr Evan J. FRS (1903–45)

A physicist who was born in rural mid-Wales, he won a scholarship to Swansea Technical College before graduating from University College of Swansea (now Swansea University). He then pursued scientific research at Swansea, Manchester, Liverpool and Aberystwyth and was awarded doctorates by the universities of Cambridge, Manchester, and Wales. In 1938 he was appointed to the chair of Physics at Aberystwyth but his tenure was interrupted by the outbreak of the Second World War. He was elected FRS in 1939.

Appointed a scientific officer at RAE, Farnborough in 1939, he worked with Patrick Blackett on the magnetic detection of submarines before moving with Blackett to apply Operational Research principles in RAF Coastal Command. When Blackett moved to the Admiralty in 1941 Williams became Coastal Command's Director of Research. In 1943 he was appointed scientific advisor to the Admiralty on methods of combating U-boats and the following year became the Royal Navy's assistant director of research. He died in 1945 at the early age of 42.

Learning that few sightings of U-boats by Coastal Command aircraft resulted in actual attacks on the submarines, Blackett argued that this was because the aircraft had been spotted too soon and so the U-boat had had time to submerge. By changing the colour of Coastal Command aircraft from grey or black to white, a dramatic increase in the number of attacks was recorded. Similarly, the success rate of aerial bombing attacks on submarines was very low until Williams produced a paper arguing that if Coastal Command flew at low level and dropped depth charges instead of bombs, at the same time adjusting the setting and spacing of the depth charges, the probability of a 'kill' would be increased by a factor of ten – which proved to be the case.

The Bay Offensive, in which U-boats were attacked crossing the Bay of Biscay between their bases and operations, was a very successful example of the effectiveness of OR. A U-boat needed to surface for some five hours a day to keep its batteries charged, but OR scientists calculated that, given sufficient aircraft, a U-boat could be forced to dive every thirty minutes to avoid being attacked. In response, U-boat commanders were ordered to submerge at night and to be prepared to defend themselves against the aircraft by day, a policy which turned out to be unsuccessful for them as the number of U-boats sunk increased.

Some Other Technical Developments

In the war's early years many other inventions aiming to improve success and survival in the maritime battle were made, particularly by the specialist Admiralty DMWD. By the summer of 1940 the Germans had realised that the British had been successful in countering the magnetic mine (see Chapter 3), and in consequence they turned to acoustic mines which were activated by the noise of a ship's engines. They were first used in November 1940, with the cruiser HMS *Galatea* being the first casualty. Fortunately, two mines were recovered intact soon afterwards and Admiralty scientists were able to dismantle them. Having ascertained how they worked the scientists then developed a means of detonating them safely. This consisted of mounting a Kango road drill in a floating box (a hammer box), two of which were then towed alongside a minesweeper and kept at a distance from the vessel by paravanes (winged devices whose wings kept them at a distance from the towing vessel). Some mines were exploded by this method more than a kilometre from the minesweeper. This technique is still in use today.

Officers and scientists from HMS *Vernon* then considered what other types of non-contact mines or other weapons might be developed so that appropriate countermeasures could be devised in advance of their being required. As a result, the impact of these weapons was very limited.

Since it would interfere with magnetic compasses, armour plating could not be used to protect men exposed on the bridge, gun platforms, or the open upper decks of merchant ships. As an alternative, concrete blocks were initially installed but these could splinter and cause heavy casualties when hit by enemy fire. After various experiments DMWD developed plastic armour, a mixture of granite chippings and bituminous mastic applied to a thin steel plate. After some internal opposition it was eventually adopted, not only for merchant vessels but for Royal Navy and US Navy ships as well, with more than 10,000 ships being treated.

This was one of the DMWD's early successes. Lieutenant-Commander E. Terrell RNVR and the Neuchatel Asphalte Co. Ltd received £9,500 and £2,000 respectively (awards No. 89 and 91) from the Royal Commission on Awards to Inventors.

Various anti-aircraft devices were considered by DMWD and others to protect merchant ships from attacks by low-flying planes. Among those which achieved varying degrees of success was the Holman Projector, a mortar-type launcher of explosive projectiles fired against aircraft, the motive force being provided by compressed air or high-pressure steam. It was successful when installed on trawlers and other small vessels, including the RN's motor gun boats. Some 4,500 were fitted by 1943. It was developed in conjunction with the Holman Company, a Cornish manufacturer of mining machinery.

Achieving limited success were: a vertical flamethrower which generated flames into the flight path of approaching aircraft; an acoustic warning device that when installed on mastheads gave advance warning of an approaching plane (it was eventually superseded by radar); and a parachute and cable device. This created a mini barrage balloon by firing steel cables whose bottom ends were attached to the vessel and their other ends to parachutes which stayed aloft long enough to deter an attacking aircraft. Initially only installed in the bows of ships, they were later fitted in the stern as well. More successful was a rocket-propelled armour penetrating spear, designed and developed to be launched from aircraft and to penetrate U-boat hulls. The device secured its first 'kill' within eight weeks of its introduction.

A problem faced by escort vessels was that U-boats, whether on the surface or submerged at periscope depth, were hard to see at night. The Navy already had a suitable flare ('Snowflake') and, by attaching it to a parachute and firing it by a rocket, DMWD developed the 'star shell' technique, which provided excellent illumination for a short time. It was also discovered that the 'star shell' had incendiary properties; it was installed without the parachute on motor gun and torpedo boats and used with devastating effect in battles with German E-boats.

Normally manned by two frogmen or up to four sailors, midget submarines were first used by the Italians in December 1941 in a raid on Alexandria harbour. Limpet mines were attached to the hulls of the battleships HMSs *Queen Elizabeth* and *Valiant* and destroyer HMS *Jarvis*, all three of which were seriously damaged. The British then developed their Chariot midget submarines (or human torpedoes) that were launched into the sea close to their target from a mother submarine. Two frogmen sitting astride the craft guided it to their target, where limpet mines were then attached to a ship's hull. Unfortunately they had little success.

More fruitful were the X-craft developed at a training base in the Isle of Bute. In these the two–four-man crew were inside the craft. In September 1943 six X-craft were towed across the North Sea so as to attack German capital ships in Norwegian harbours. Of the six only two were able to complete their mission, but they successfully laid time-fused mines on the seabed under the German battleship *Tirpitz* and seriously damaged it. X-craft were also used to reconnoitre the Normandy beaches in advance of the D-Day landings.

First investigated by Blackett and Williams when they were working at the Royal Aeronautical Establishment, the magnetic detection of submarines was based on the ability of an instrument in a low-flying aircraft to recognise a deviation in the earth's magnetic field when a submarine was below it. The idea was conveyed to the Americans, and as a result, a magnetic anomaly detector (MAD) was developed. This was fitted to many aircraft and achieved a number of successes. Towards the end of the war MAD was largely supplanted by the cheaper radio sonobuoy (although both are still in use today).

Sea Forts

To counter enemy aircraft approaching London or Liverpool from the sea, a number of sea forts were constructed in isolated open water positions near the Thames and Mersey estuaries. They could give early warning of the approach of hostile aircraft or naval forces, as well as deter the laying of mines in navigable channels. In addition to the normal hazards of distance, tide and tempest, they were vulnerable to attacks by the *Luftwaffe* and enemy surface and submarine craft.

Altogether ten forts were built, seven in the Thames Estuary and three in the Mersey Estuary, manned either by naval personnel or Army units. Designed by civilian consultants to Admiralty and War Office requirements, the forts were nicknamed 'Maunsell forts' after their chief designer G. A. Maunsell. Some of the forts comprised a

The Maunsell Sea Forts of WW2. Neil Brown

Naval fort ready for action. Richard Brown

Sinking the *Bismarck* – with old technology triumphant

The sinking of the imposing 52,600-ton German battleship *Bismarck* in May 1941 was one of the war's turning points. Admiral Tovey, the commander in chief, graciously commented that 'She put up a most gallant fight against impossible odds, worthy of the old days of the Imperial Germany Navy,' but Admiral Raeder was blunter in his analysis: 'The loss of the *Bismarck* had a decisive effect on the conduct of war at sea.'

Impressive though advances in Allied technology had become, the truth is that on this occasion it was old technology – a combination of ingenious tactics, experienced aircraft and vessels, and conventional weaponry – that ultimately defeated her.

It had long been the Admiralty's fear that a German capita] ship would break out from its home port and wreak havoc in the North Atlantic and in May 1941 the *Bismarck*, the world's most powerful ship, and the heavy cruiser *Prinz Eugen*, sailed from the Baltic to Norway and then towards the Denmark Strait (between Greenland and Iceland).

The ships were first shadowed by RAF Coastal Command and RN Fleet Air Arm aircraft, but after being lost in the Arctic mists, they were located again in the Denmark Strait thanks to a Bletchley Park decrypt. They were then engaged by HMSs *Hood* and *Prince of Wales*, but one of the *Bismarck*'s first salvos penetrated the *Hood*'s relatively thin upper deck armour and entered the magazine, thereby causing a major explosion which resulted in the disintegration of the ship with the loss of all but three of the 1,400-strong ship's company.

The *Prince of Wales* broke off the engagement and then shadowed the *Bismarck*, which had decided to sail south-eastwards for the safety of Brest. (She had suffered some relatively minor damage and had not been fully refuelled when in Norway.) Meanwhile the RN brought all available ships to the chase, including the new battleship HMS *King George V*, the old battleships HMSs *Revenge* and *Rodney*, and the Gibraltar-based *Force H*, which comprised the battlecruiser HMS *Renown*, aircraft carrier HMS *Ark Royal*, and cruiser HMS *Sheffield*.

Although shadowed by Coastal Command aircraft and radar-carrying RN cruisers, the *Bismarck* was temporarily able to give her pursuers the slip in the Atlantic mists. However, after radar contact had been regained, open cockpit Swordfish biplanes, which had a top speed of only 224 km/h (140 mph), were launched from the carrier HMS *Victorious* and attacked with torpedoes. They achieved one hit, but the *Bismarck*'s thick armour plating made it largely ineffectual.

During the following night radar contact was again lost when the shadowing cruiser was undertaking an antisubmarine zig-zag and *Bismarck* was simultaneously changing her course. Amazingly, the RN hunters now began to run short of fuel and, rather than go to Iceland to refuel, they were forced to reduce speed. This left only *Force H* in the main chase, and its ships were fast approaching *Bismarck from* the southeast.

Bismarck was eventually located again on the morning of May 26 by a Coastal Command Catalina aircraft (whose co-pilot was a USN ensign, despite the Americans not having entered the war). Despite the weather causing her deck to rise and fall some 17 metres, an attack was launched *by Ark Royal's* open-cockpit Swordfish biplanes, their torpedoes primed with a new magnetic pistol.

The airmen were given a bearing on which to find the enemy, and when a ship was sighted through a break in the clouds they launched their torpedoes. Unfortunately, this was HMS *Sheffield* and not the *Bismarck; a* breakdown in communications having resulted in the aircrews not being informed that *Sheffield* was now shadowing the *Bismarck. Sheffield* held her fire, and by a combination of good fortune and good seamanship, survived intact. Five of the magnetic pistols malfunctioned and the ship managed to steer clear of the other six torpedoes that were fired towards her.

A second attack began later that day but the torpedoes were now armed with the older contact pistols. Thirteen torpedoes were launched and two, possibly three, hit *Bismarck.* Of these two hit her amidships and caused only slight damage but one was well aft and incapacitated the ship's rudders, thereby causing her to go round in circles. Despite valiant efforts by her divers and damage-control parties it proved impossible to move the rudders and attempts by the captain to navigate on engines also proved ineffective. The attack by the Swordfish aircraft had lasted for just half an hour but this was one of the most decisive half hours in the history of naval warfare.

The British capital ships re-entered the battle the following morning and by 10am the *Bismarck* was afire and a floating wreck and the cruiser HMS *Dorsetshire* finished her off with torpedoes. Of *Bismarck*'s company of 2,200 there were only 110 survivors.

Torpedo-carrying Swordfish airplane. Tony Hisgett

number of towers interconnected by walkways. They were built in nearby harbours and then towed out to their action stations – a precursor of post-war oil platforms.

By these various means the Allied navies eventually gained the upper hand over the U-boats and, although February 1943 had seen some of the highest Allied shipping losses of the whole war, in the summer of that year, the U-boats were withdrawn from the Atlantic. And, although they reappeared later, they were never again the threat that they had been in the early years of the war.

Avoiding defeat in the Battle of the Atlantic enabled the Anglo-American allies to continue the fight in Europe, mount offensive actions in North Africa, Italy, and north-west Europe, and support Russia by supplying her with armaments via her Arctic Circle ports...

RUSSIA TIMELINE

1925–26

Hitler publishes his two-volume book *Mein Kampf* (My Struggle) in which he highlights his detestation of Marxist Russia (and of the Jews) and the need for Germany to expand eastwards so she can acquire more *Lebensraum* (living space).

1939

31 Mar UK and France give guarantee to Poland after German annexation of Czechoslovakia.
31 Aug Russian-German Non-aggression Pact announced.
1 Sep Germany invades Poland.
3 Sep UK and France declare war on Germany.
17 Sep Russia invades eastern Poland.
Sep Warsaw surrenders. Poland wholly occupied by Germany and Russia.

1940

Russia massacres c.22,000 Polish army officers, professionals and civil servants.

1941

22 Jun Germany invades Russia. Britain pledges support.
Aug Germans launch attacks on Moscow and the Crimea. Russia retakes Rostov.
Britain and Russia enter Iran to secure oilfields and provide Russian-Persian Gulf link.
First Arctic convoy sails.
Sep Germans occupy Kiev; Siege of Leningrad (now St Petersburg) begins.
Oct Siege of Moscow begins.
7 Dec Germany declares war on USA, who then supplies and participates in Arctic convoys arms to Russia.

1942

Jan Siege of Moscow lifted.
Russian counter-offensive into the Ukraine.
May German offensive in Kerch Peninsula.
Jul Arctic convoy PQ 17 sails. Only 11 ships (out of 35) complete the trip.
Aug Siege of Stalingrad begins.

1943

Feb Germans surrender at Stalingrad and siege lifted. Regarded as the Russian campaign's 'turning point'; Hitler reputedly a 'broken man' afterwards.
Jul Germans launch new offensives at Kursk, Orel, and elsewhere.
12 Jul Largest tank battle (with 800 Russian tanks) fought at Prokhorovka, Kursk.

1944

Jan Siege of Leningrad raised.
Jun–Aug Russia's Operation Bagration in Belarus overwhelms central German Army.

1945

Feb Russian troops cross River Oder and enter Germany.
23 Apr Russian troops reach Berlin suburbs.
30 Apr Hitler commits suicide.
12 May 75th (and last) Arctic convoy sails.
2 May Berlin surrenders.
7 May Germany surrenders unconditionally.
8 May VE Day.

Chapter 7

THE WORST JOURNEY IN THE WORLD

Russian defiance and German capitulation

The magnitude of the Russian campaign fought between Germany and the Soviet Union is frequently under-appreciated in the West. The bloodiest of the whole war, it claimed the lives of some 30 million people due to combat, starvation, exposure, disease and massacres.

One aspect of it, however, is firmly fixed in British consciousness – the heroic Anglo-American supply convoys to the Russian Arctic ports of Murmansk and Archangel. Churchill memorably called their dogged progress through icy conditions and under heavy bombardment 'the worst journey in the world'. These convoys carried tanks, crated aircraft, munitions, and other impedimenta. As was later acknowledged by Stalin, they made a massive contribution to the Russian success in winning the eastern land battles.

Escorted mainly by RN ships (and also some RCN and USN vessels) the ships sailed from British and Icelandic ports and were vulnerable to attacks by German air and sea forces based in Norway. These included a number of *Luftwaffe* bases and the battleship *Tirpitz* – sister ship to the *Bismarck* – as well as battlecruisers and pocket battleships. In addition they faced Arctic pack ice and storms, perpetual night in winter and perpetual light in summer, unreliable compass readings because of the high latitudes, and icing on ships' superstructures.

The ships' engineers had to ensure that the accumulation of ice on a vessel's superstructure did not adversely affect its metacentric height to such an extent that she would be more likely to capsize. In some cases, propellers could be chipped or otherwise damaged by striking ice, resulting in additional effort being needed to ensure the ship maintained its position in the convoy or with the escorts. If a ship should be hit by the enemy, its damage-control parties had to fight fires, plug holes, and provide emergency support for stricken bulkheads and deckheads.

From August 1941 until the end of the war 40 outward convoys sailed comprising 811 merchantmen and carrying over 4 million tons of supplies, including 5,000 tanks and over 7,000 aircraft. There were 35 return convoys with 717 merchant vessels.

In total 87 merchant ships were sunk. A few had to return to harbour for various reasons, but the vast majority completed the journey despite all the hazards. In the

Enter a Hero

The Battle of the Barents Sea was fought north of North Cape on 31 December 1942 between British and German naval forces; each side lost a destroyer and had one ship damaged. The British flagship, the destroyer HMS *Onslow*, was badly damaged causing, among other things, a fire in the engine room, but the ship was saved thanks to its engineer officer, Lieutenant (E) Kevin Walton RN. He made and donned a rope harness in which he was lowered into the engine room with a hosepipe and he managed to extinguish the flames – a feat for which he was awarded the DSC.

Walton, Lieutenant (E) Kevin GC DSC MICE (1918–2009)

He was born in Japan where his father was a missionary. Thanks to a godfather who had been on two Everest expeditions, he acquired an interest in mountaineering and, through that, knowledge of knots. Bizarrely, this resulted in him being awarded two gallantry medals during and immediately after the war. He read civil engineering at Imperial College and served in the Second World War as an engineer officer in the Royal Navy. In the Battle of the Barents Sea in December 1942 his ship, HMS *Onslow*, was damaged and the ship's engine room set on fire. Undaunted, he was lowered on a rope harness into the engine room with a hosepipe and extinguished the fire, thereby saving his ship. His bravery was recognised by the award of the DSC. Having expressed an interest in Antarctica, he was rapidly demobilised at the end of the war so that he could join the first British post-war expedition to Antarctica. When a colleague inadvertently fell into a crevasse and became trapped Walton volunteered to be lowered alongside him. By chipping at the ice he was able to free his colleague, for which he was awarded the Albert Medal (subsequently converted to a George Cross). Among his later occupations were: teaching at various schools and the Royal Naval College, Dartmouth; pioneering sailing courses in Scotland; and initiating a system to engage schoolchildren in engineering. This has evolved into STEMNET which has nearly 20,000 STEM (science, technology, engineering, and mathematics) Ambassadors. He also served on the council of the Institution of Civil Engineers.

On Board *Bermuda*

On 26 December 1943, the Battle of North Cape was fought resulting in the sinking of the German battlecruiser *Scharnhorst* with the British only suffering minor damage to three ships. Among the naval ships that escorted the convoys was HMS *Bermuda* one of whose engineer officers, Lieutenant (E) Teddie Drew RN, wrote a post-war account of his naval service which can be found online.

Drew, Lieutenant (E) Edwin 'Teddie' FICE RN (1917–2015)

Drew joined HMS *President*, the London-based training establishment of the Royal Naval Volunteer Reserve, as a communications rating in 1937. Having obtained an engineering degree at University College London, he was given an engineering commission shortly after the war started. After his ship, HMS *Cornwall*, was sunk by Japanese bombers in the Indian Ocean he was in an open boat for 24 hours. His next ship, HMS *Bermuda*, did a number of Arctic convoy runs; he described much of his adventures online. He was presented with his Arctic Star at a ceremony in HMS *President* in December 2013. Post-war he became manager of a major drainage authority in Hertfordshire and served as president of the Institution of Public Health Engineers.

battles the RN lost two cruisers, six destroyers, and eight escorts while German losses included one battlecruiser, three destroyers, and more than 30 U-boats, as well as a large number of aircraft.

The Persian Corridor

The Arctic wastes weren't the only supply route to Russia. Shortly after the German attack on Russia, Anglo-American and Russian forces occupied Persia (now Iran) in order to secure Persian oil supplies and provide a transportation link to convey western armaments to Russia.

Known as the Persian Corridor, it comprised rail and road links from the Persian Gulf ports of Bandar Shahpur (now Bandar-e Emam Khomeyni) in Iran and from Basra in Iraq to Teheran and thence on to Azerbaijan, one of the southern Soviet states. Some Caucasian oil was also transported in reverse.

The single-track standard gauge railway running from the Persian Gulf to the Caspian Sea was upgraded by British, American, and Russian engineers so as to be able to run 12 to 15 trains per day carrying a total of some 5,400 tons/day of aid. Western engineers assumed responsibility for the operations south of Teheran while Russian

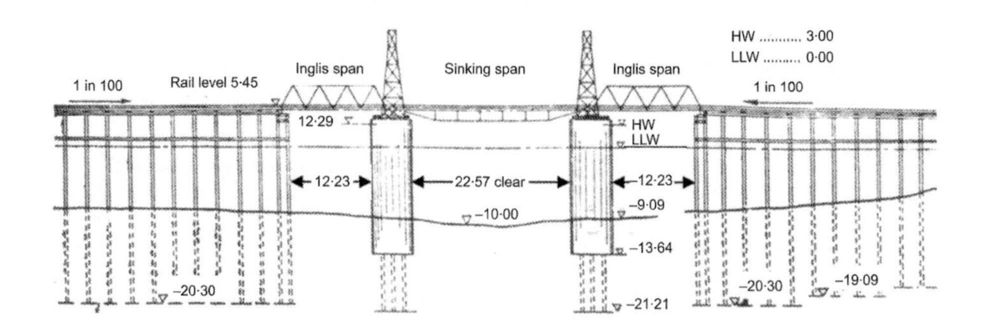

Hull Bridge over the Shatt al-Arab River at Basra, Iraq. ICE

engineers took charge in the north. Both sections were difficult because of the nature of the terrain, and it was a mammoth task as the railway passed through gorges, bridges, and tunnels and rose to over 2,000 metres before descending again and had gradients of up to 2.3%. It has been described as comparable to 'some of the mighty railway achievements in the Rockies in Canada and the USA'.

Goods landed at the Iraqi port of Basra had to be conveyed across the wide Shatt al-Arab River in order to access the Persian transportation network. To achieve this, sappers from the British and Indian armies designed and constructed an ingenious bridge which thoroughly demonstrated the resourcefulness of engineers when required to find an immediate solution to a pressing problem.

As it was necessary to maintain river transport as well, and as the facilities in the local RE workshops were unable to produce the hoisting equipment needed for a central lifting span, it was decided to design and build a central dropping span which could be lowered into the river deep enough to be below the keels of the river vessels. Known as the Hull Bridge after the RE officer in charge of the operation, it was 600 metres long, including approach viaducts supported on 10-metre-long wooden piles that were surplus to requirements following the construction of a deep-water jetty nearby.

The Pacific and Other Routes

Although Japan was at war with the USA from December 1941, Japan and Russia were not with each other and both observed strict non-belligerency conditions against the other. As a result, certain non-military supplies were sent across the North Pacific Ocean in Russian-flagged ships from US West Coast ports to the port of Vladivostok in the Russian Far East. The non-belligerency arrangements restricted these to supplies other than war materials, but they included lorries, cars, railway engines, and wagons. It has been said that about 50% of all US supplies to Russia were transported by this route.

As well as these, some 8,000 US-built combat aircraft which were manned by Russian crews were flown directly from Alaska to Siberia. Some lighter aircraft were crated and shipped to Vladivostok, and thence westwards along the Trans-Siberian railway.

Crated British Hurricane aircraft were sent on some of the early Arctic convoys, while long-range American aircraft were flown from Florida to Brazil, then to the west coast of Africa, along North African coasts to the Middle East, and from there to Abadan or Tehran, where Russian crews took over. Some lighter aircraft shipped to Basra were reassembled before being flown to Tehran.

The Land and Air War

While the achievements of Western engineers in supporting the Russians were significant, those of the Russian engineers were even more outstanding. Sadly, comparatively little has been written about them compared with the voluminous information available about the technical details of the Western Allies' war effort. Among recent accounts of the Russian campaign has been Paul Kennedy's book *Engineers of Victory*, and much of the following description has been derived from it and from other sources online.

Red Army engineers made major contributions to achieving victory in the various battles that were fought. Minefields up to 40 km deep were laid, thereby effectively eliminating the prospect of lightning strikes by German armour. They also successfully constructed crossings of the main rivers for their armies.

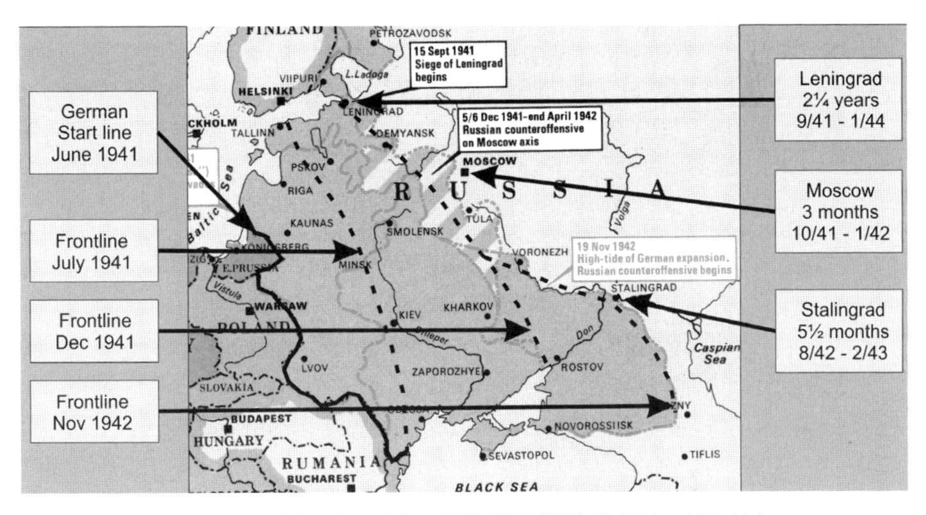

German advances and sieges. Map from Atlas of WORLD WAR II, Richard Natkiel

Russian engineers not only manufactured vast quantities of armaments but also produced some excellent designs that allowed their armoured vehicles, artillery, and aircraft to perform very creditably against the much vaunted German equivalents.

This is, of course, praiseworthy in itself, and it is all the more remarkable since, as a result of the German advances, many factories in western Russia had to be dismantled, transferred eastwards and rebuilt in the Urals or Siberia. For instance, the Voroshilov tank factory moved from Leningrad to Omsk, a distance of nearly 2,000 km.

Throughout the campaign the Russian T-34 tank held its own against German armoured vehicles. It was faster, more manoeuvrable, and with better range and firepower; the latest versions operating in the last year of the war were considered superior to any other armoured vehicles produced by any of the belligerent nations.

Other very useful Russian weapons included the multi-headed anti-tank *Katyusha* rocket launchers and the PaK bazookas, the infantry-operated anti-tank weapon. Russian production engineers were highly efficient – the Russians were able to deploy more than 20,000 guns and mortars in the Battle of Kursk. Anti-tank and anti-personnel mines were created in very large numbers too.

Unlike the Anglo-American air forces which also had to participate in the Battle of the Atlantic and mount the strategic bombing campaign against Germany, the Russian Air Force could concentrate its efforts on fighters and in providing tactical support to the Red Army.

Its principal fighter, the *Yakovlev*-3, (Yak-3) was claimed by the Russians to be as good as contemporary Spitfires, but as they were never engaged in similar actions it is difficult to make a comparison. Tactical support aircraft included the *Ilyushin Sturmovik*, which proved to be an effective anti-tank aircraft.

In the pre-First World War Tsarist regime American and other Western aeronautical companies established links with Russia, although these were mostly discontinued after the Revolution and not revived until the late 1930s (although the Russian Navy purchased Italian flying boats in 1929). With a thawing of the Western attitude towards the Communist regime, the Russians were able to purchase single aircraft which were

T-34 tank. http://www.tanks-encyclopedia.com/ww2/soviet/soviet_T34-76.php

Ilyushin Sturmovik airplane. Flying Heritage Collection Paul Allen Microsoft

studied by their designers when preparing plans for new aircraft. Following the entry of Russia and America in the war in 1941, vast numbers of American aircraft were supplied under lend-lease arrangements. In April 1944 there were more than 13,000 Russian aircraft opposing the *Luftwaffe*, many of which must have been built in Russia itself.

Anglo-American Allies Divert German Attacks

The early campaigns in North Africa, Italy and other Mediterranean countries, and later in north-west Europe, all resulted in the Germans having to fight on more than one front, thereby diluting the forces available to attack Russia. Furthermore, the increasingly heavy bomber offensive on German cities meant that elements of anti-aircraft artillery and the *Luftwaffe* had to be used to defend the homeland, thereby decreasing the numbers available to fight against Russia. Similarly, the continuing German efforts to win the Battle of the Atlantic diverted the scarce supply of materials and skilled workers into constructing U-boats etc., thus reducing the weapons and other armaments available for the Eastern Front.

Stalin, in his 1944 May Day speech, praised the Western Allies for diverting German resources to the Italian Campaign, while the Russian news agency *Tass* published detailed lists of the large numbers of supplies coming from the Western Allies. Six months later Stalin made a speech stating that Allied efforts in the West had already drawn 75 German divisions to defend that region, and without this diversion the Red Army could not yet have driven the *Wehrmacht* from Soviet territories. Russian engineers were therefore indirectly assisted in the defeat of Germany in the East by their Western counterparts.

The victories in the East coincided with Anglo-American victories in North Africa, Italy, and north-west Europe...

MEDITERRANEAN TIMELINE

1939

7 Apr	Italy invades Albania.

1940

11 Jun	Italy declares war on Britain and France.
3 Jul	RN attacks French battleships in Algiers harbour and neutralises others in Alexandria.
4 Jul	Italians start attacks on Abyssinia (now Ethiopia) and British colonies in Horn of Africa.
Sep	Italian troops in their Libyan colony advance into Egypt (where UK has bases).
28 Oct	Italian forces in Albania invade Greece.
11 Nov	RN aircraft successfully attack Italian fleet in Taranto Harbour.
9 Dec	Egyptian-based UK troops advance into Libya; see-sawing campaigns follow.

1941

Jan–Apr	British Army achieves victories in Libya, Abyssinia, and Horn of Africa.
Mar	RN victorious in Battle of Cape Matapan; Italians counter-attack UK forces in Libya; German *Afrika Korps* formed to support Libyan-based Italians.
3 Apr	Pro-Axis *coup d'état* in Iraq overthrows pro-British Iraqi Government.
6 Apr	Germany invades Greece and Yugoslavia.
22 Apr–2 May	UK evacuates Greece.
20 May	Germany invades Crete.
30 May	Iraqi revolt collapses.
1 Jun	UK evacuates Crete.
25 Aug	UK and Russia enter Iran.
10 Dec	USA enters war.

1942

1 Jul	After see-sawing Libyan/Egyptian campaigns, UK stands at El Alamein.
3 Nov	UK victorious in Battle of El Alamein; Axis forces then withdraw into Libya.
7 Nov	UK and US armies land in Morocco and Algeria; German troops fly into Tunisia.

1943

Feb–May 12	Germans retreat into Tunisia where land battles result in German defeat.
10 Jul	Allies invade Sicily.
17 Aug	Germans evacuate Sicily.
3 Sep	Allies invade Italian mainland.
8 Sep	Italy surrenders but Germans augment their forces and continue fighting.
1 Oct	Allies take Naples.
1 Dec	German winter line broken.

1944

16 Jan	Anzio landings.
4 Jun	Rome captured.
6 Jun	D-Day; US/UK/Canada invade Normandy.
15 Aug	US and UK invade the south of France.

1945

28 Apr	Mussolini captured and shot by partisans.
2 May	Unconditional surrender by Germans in Italy.

Chapter 8

THE END OF THE BEGINNING

North Africa, the Mediterranean and Italy

As usual, Churchill found the words for it. The Allied victory at El Alamein – securing the Suez Canal, Palestine, Syria, and the Persian oil fields – was by no means the end of the war, and not even the beginning of the end, but it was 'the end of the beginning'. That was in November 1942, the exact midpoint of the campaigns in North Africa, the Mediterranean, and Italy, in which the engineering fraternity was called upon to make a series of significant interventions – not least among them the laying and clearing of minefields.

In the August of that year General Bernard Montgomery, newly appointed commander of the British Eighth Army, had faced a German offensive at Alam Halfa. Aided by daily Bletchley Park decrypts, Montgomery knew of Field-Marshal Erwin Rommel's plans, tank strengths, and fuel stocks and was able to repel the attack over the ensuing week-long battle.

Prior to the start of the battle the British had laid extensive minefields and these, coupled with large numbers of anti-tank artillery, were largely responsible for resisting the German thrust. Lanes which had been left through the minefields for British patrols had to be closed when the Germans attacked, and Sapper Kenneth Stansfield was awarded the DSM for his gallantry in completing this task under close-range fire from an enemy tank.

After repulsing the German onslaught and having completed the build-up of his army, Montgomery prepared to launch his own attack from his position at El Alamein. For it to be successful, however, it would be necessary for his tanks to advance through enemy minefields. Tanks had been invented in 1916 as a means of breaking the deadlock of the trench warfare of the First World War and as a solution to the slaughter of the infantry, the theory being that troops in armour-protected vehicles could lead an advance with the infantry following behind. A quarter of a century later there was a complete U-turn, it having been realised that while tanks could be effective in open terrain, they were very vulnerable to damage and destruction if trying to advance through a well-prepared minefield.

This was the situation faced by the British Eighth Army when preparing to advance from its defensive position at El Alamein. The solution was for the sappers to clear lanes through the minefields so that the tank formations could follow. Traditionally, enemy

Minefield clearance at El Alamein Painting by Terence Cuneo (by permission of the Institution of Royal Engineers)

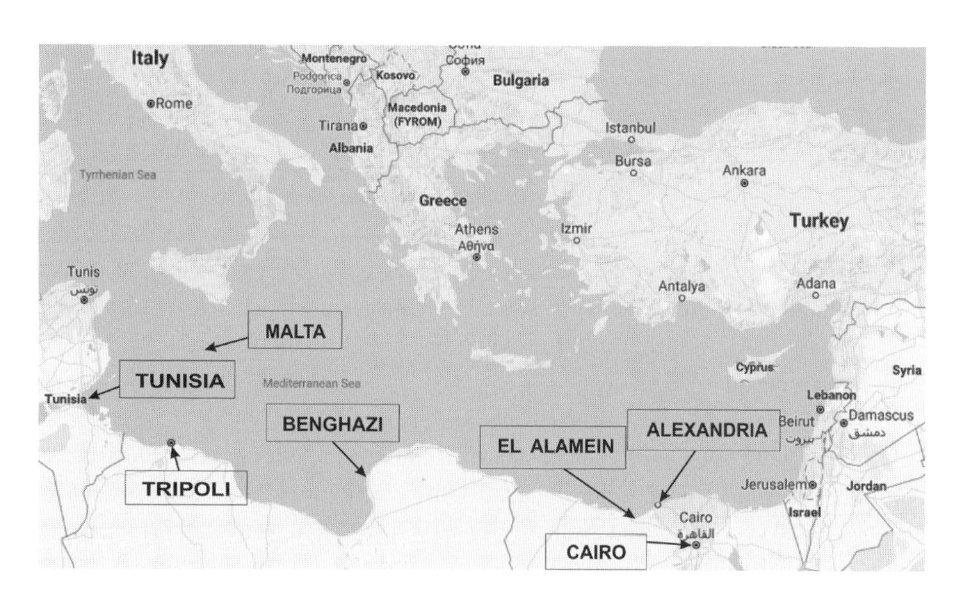

mines had been located by sappers probing the ground with handheld bayonets, but newly developed handheld mine detectors were now used for the first time. These had been developed in Scotland by Lieutenant Józef Kosacki, a Polish engineer officer. These doubled the speed at which heavily mined sands could be cleared, from an advance of around 100 metres an hour to about 200 metres an hour. One RE squadron was also assigned to accompany an armoured division as it advanced so as to clear enemy minefields.

Kosacki, Professor Józef Stanislaw (1909–1990)

An inventor and officer in the Polish Army in the Second World War, he created the first man-portable mine detector, whose basic design has been in use with various armies for over 50 years. Before the Second World War he had been a technician in the Artillery Department of the Polish Ministry of National Defence. Shortly before the war he joined the clandestine Special Signals Unit, a secret organisation that worked on electronic appliances for the Army. Following the 1939 invasion of Poland he managed to get to the United Kingdom, where he continued his service in the Polish Army as a signals officer. In 1941 he devised a mine detector that was used in action for the first time in the Battle of El Alamein, the Eighth Army having received 500 of the detectors. This doubled the speed at which heavily mined sands could be cleared, from 100 to 200 metres an hour. During the war his name was classified in order to protect his family, who had remained in German-held Poland. Therefore most of his patents were submitted under pseudonyms, including 'Józef Kos', 'Kozacki', and 'Kozak'. As a result, his surname is often given erroneously in post-war historiography. After the war he developed electronic and nuclear machinery and was a professor at the Institute for Nuclear Research at Otwock-Świerk as well as at the Military Technical Academy in Warsaw. He died in 1990 and was buried with military honours. In 2005 the Wrocław-based Military Institute for Engineering Technology (WITI) was named after him. This institute has the first prototype Polish mine detector built by Kosacki.

The ensuing British victory was brought about by a considerably greater strength in the numbers of men, armour, and aircraft available to Montgomery; knowledge about depleting German fuel stocks gleaned from Bletchley Park decrypts; and improved communications between individual army units and tactical support aircraft, as well as other new technology being deployed which made British weapons more effective.

Allied Landings in the Mahgreb

On 8 November 1942 Anglo-American forces made largely unopposed landings in the French Maghreb territories of Morocco and Algeria, while the Germans flew reinforcements into Tunisia. Over the next few weeks Rommel's *Afrika Korps*, chased and harried by Montgomery's Eighth Army, retreated westwards into Tunisia where fierce fighting ensued as they fought Allied forces on two fronts. This eventually led to the surrender of the Axis forces on 12 May. The fighting in Tunisia saw the first involvement of the Bailey bridge and of RE parachute troops, an integral component of the Parachute Brigade. Both of these grew in importance in the later phases of the war.

Abyssinia and the Horn of Africa

Operations in North Africa started in the summer of 1940 after Italy entered the war. Some of the early actions involved fierce fighting which led to the eventual liberation of Abyssinia (now Ethiopia) and the former Italian colonies in the Horn of Africa, now Eritrea and part of Somalia.

A Significant VC

During the Eritrean campaign an Indian Army engineer, Lieutenant Preminda Singh Bhagat, was awarded the VC, part of his citation reading:

> For a period of four days and over a distance of 55 miles this officer in the leading carrier led the column. He detected and supervised the clearing of fifteen minefields. Speed being essential he worked at high pressure from dawn to dusk each day. On two occasions when his carrier was blown up with casualties to others, and on a third occasion when ambushed and under close enemy fire, he himself carried straight on with his task. He refused relief when worn out with strain and fatigue and with one eardrum punctured by an explosion, on the grounds that he was now better qualified to continue his task to the end.

This was the first VC won during the Second World War by an engineer or by a member of the Indian Army.

Cardboard Tanks

The campaigns in Egypt and Libya, which also began in the summer of 1940 and continued for the next three years, were confined to a narrow coastal strip along the Mediterranean and presented some difficult and unexpected engineering challenges. Not least amongst them were the consequences of the see-sawing campaigns; these involved the rapid movement of the front, which advanced or retreated some 3,000 km at an average rate of some 16 km/day. The logistics of providing supplies of fuel, water, food and munitions was always a major consideration, and particularly so when the front line was advancing so rapidly. When it was far removed from Alexandria, supplies to forward positions often had to be delivered by sea or air.

In addition to the provision of accommodation, water etc. for about 130,000 prisoners of war, an unexpected task for the RE was the construction of dummy tanks made of cardboard for deception purposes. One brigadier commented that his life's ambition had been to command a brigade in action but that he had not expected it 'to have been built by the Sappers from cardboard'. (He was nevertheless victorious and took nearly 10,000 prisoners.) The RE also constructed defensive structures at Tobruk, which was besieged from April to December 1941 following a counter-offensive by the enemy.

The need to route sea-borne supplies and troop reinforcements around South Africa after Italy's entry into the war resulted in transportation times of some three months, which added to the difficulties of all three services in obtaining reinforcements, equipment and spare parts.

Among the RE's other major tasks during the campaigns in the Western Desert were the construction of advanced airfields for the RAF and the supply of water to the Army's 220,000 troops and to its thousands of armoured and transport vehicles, all of whose radiators had to have freshwater. To achieve this, suitable sources had to be found and operated and pipelines had to be constructed at amazing speed.

Greek Tragedy

RE units were among the troops sent to Greece in the summer of 1941 for the short-lived attempt to resist the German invasion. RE operations there included destroying the bridge over the Corinth Canal during the retreat on the mainland, and soon afterwards having to fight as infantry in Crete because of the lack of equipment.

From 1943 onwards the British supported Greek and Yugoslav resistance efforts through arms drops, clandestine operations, and missions to support the partisans. Among the engineers involved in such operations was Captain Kenneth Scott RE who, as an SOE officer, spent 18 months in occupied Greece undertaking sabotage and demolition projects.

Scott was parachuted into Greece as part of an SOE sabotage team in 1943. After successfully demolishing a railway bridge, he was instructed to destroy the heavily guarded 101-metre-high, 122-metre-span Asopos railway viaduct in southern Greece so as to interrupt the movement of German supplies before the Allied invasion of Sicily. His first

Scott, Major Kenneth MC* FREng FICE RE (1918–2007)

He served with the RE during the Second World War, going to France with the BEF in 1939 and later spending 18 months on clandestine operations in occupied Greece, for which he was awarded the MC and bar. Finally, he was sent by SOE to Thailand. After demobilisation he joined Sir Alexander Gibb & Partners (now Jacobs) in 1946 and worked on projects in New Zealand and Scotland before being appointed a partner in 1959 and senior partner in 1977. He was vice-president of the Institution of Civil Engineers in the 1980s.

attempt had to be aborted because of changed circumstances which necessitated contacting SOE headquarters in Cairo to request the dropping of additional equipment.

On the second attempt, having approached at night by a hazardous route along the bottom of the gorge, Scott found that the viaduct was under maintenance and some scaffolding had been erected, which facilitated his task. He placed and set the demolition charges and beat a hasty retreat – and was delighted to hear the explosions a couple of hours later. He subsequently recounted: 'We fixed our charges at night and waited for the first train to pass. By luck we got an ammunition train. The thing went on, a pyrotechnic display, for about four or five hours.' Scott was awarded the first of his two MCs for this operation.

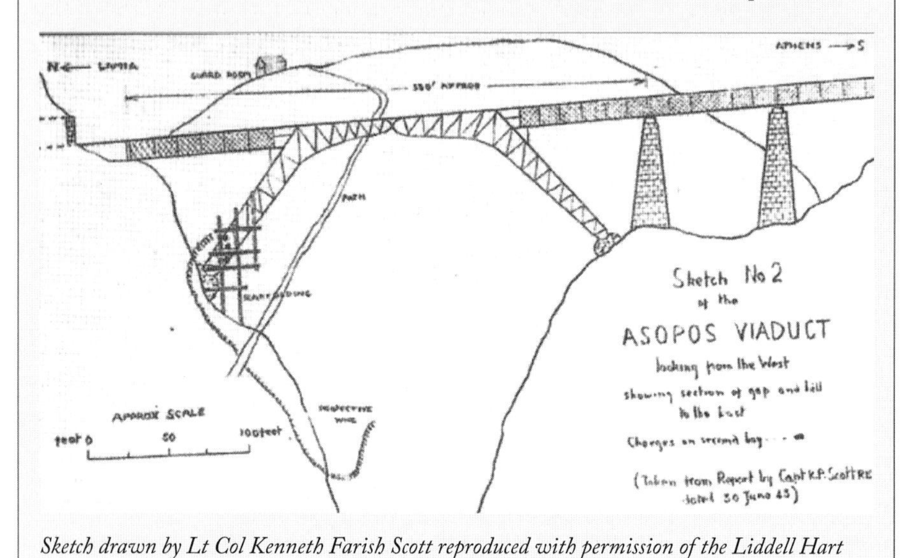

Sketch drawn by Lt Col Kenneth Farish Scott reproduced with permission of the Liddell Hart Centre for Military Archives.

RSigs, RAOC, and REME

Together with the RE, the RSigs, RAOC/REME and RN all provided tactical, technical, and logistical support for the Army's fighting units for all these campaigns, which were invariably in situations where indigenous resources and facilities were either non-existent or very sparse.

Until 1 October 1942, when REME was established, the Royal Army Ordnance Corps (RAOC) was responsible for the maintenance and repair activities that later became the responsibility of REME. Base workshops employing over 20,000 personnel were established in the Nile Delta. Field workshops and Light Aid Detachments undertook immediate recovery and repair near the battle areas, but these operations were complicated by the frequent rapid movement of the front line.

Can You Hear Me?

The swift progress of the front line caused unprecedented problems for Royal Signals; for this and other reasons radio communications had to be used to a greater extent than anticipated. Moreover, the need for effective communications with the RAF was highlighted at an early stage.

In the advance from the Western Desert the RSigs continued with their earlier activities, although in the First Army's operations in the Maghreb they had to collaborate closely with their American and French counterparts.

A joint landline communications system was established, largely based on the existing French network which connected Casablanca in the west with Allied Force Headquarters at Algiers, the various service operational headquarters and the forward elements further east. Submarine cables were picked up and diverted into Algiers so as to provide links with London, Washington and Gibraltar. High-speed radio circuits also provided communication links between the various international headquarters. For much of these developments the British and American signals units were effectively woven into one organisation.

Lessons learned included: all equipment must be 100% operationally ready at the start of a battle; repairs could not be carried out without wireless communications and an efficient recovery organisation; and the outcome of the battle could depend on the number of tanks that could be repaired and prepared for action for the following day.

Air Operations

In both North Africa and the Mediterranean the RAF's overriding concerns were the conservation of their scarce resources of men and machines and the constant presence of sand, dust and heat, which aggravated flying conditions and maintenance. Dust frequently drifted like snow, covering everything with a fine, very soft, yellow powder that reduced visibility to a few metres.

Air cleaners had to be serviced every five hours, a task which took three hours of maintenance. Sand penetrated into the instruments and interfered with the variable pitch propellers, while the high temperatures melted the plastic that secured the Perspex window sheets which then blew out or cracked.

When Italy entered the war Air Marshal Sir Arthur Longmore, the commander in chief, had none of the latest aircraft among his resources – his most modern fighter, for instance, being the Gloster Gladiator biplane which had entered service in 1937. Fortunately, the Italian opposition was of comparable vintage and in some regards amazingly ill-prepared; many of the aircraft had no sand filters. In the event the RAF personnel and aircraft acquitted themselves extremely well and outfought the Italian Air Force.

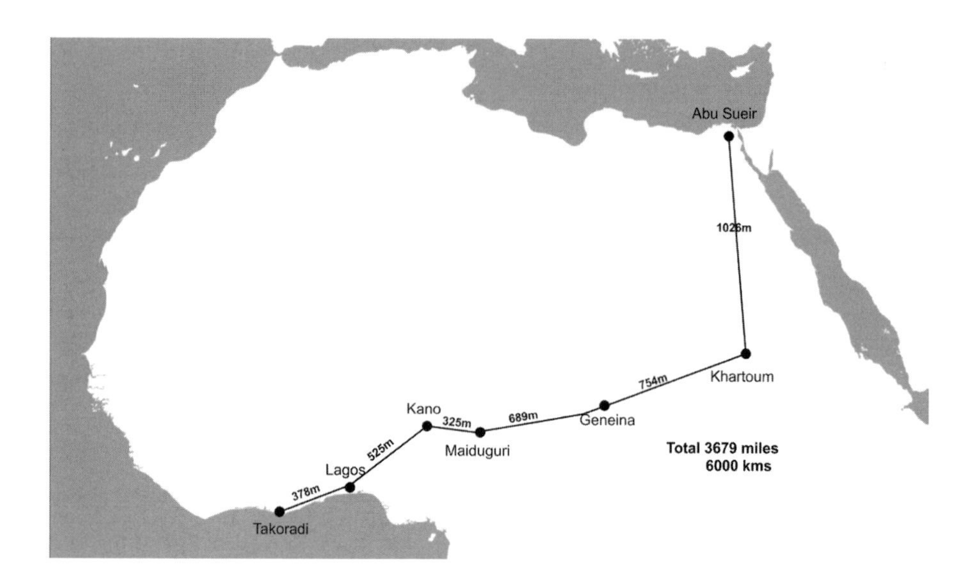

Furthermore, Italy's central position in the Mediterranean meant that air supplies and reinforcements for the British forces in Egypt could no longer go through the Mediterranean directly from Gibraltar. As a result, in July 1940 Group Captain Henry Thorold RAF and a group of technicians arrived in Takoradi in the Gold Coast (now Ghana) to develop the circuitous 'Takoradi route'. Aircraft flew from Takoradi via Nigeria and an airfield in French North Africa held by General de Gaulle's Free French forces to Khartoum in Sudan and thence north along the valley of the Nile to Egypt. Later described as 'a first-class piece of improvisation', the first transit of six Hurricanes and one Blenheim aircraft left Takoradi on 20 September 1940. The operation of the route eventually absorbed nearly 70,000 RAF personnel; by 1943 it had enabled some 5,000 aircraft to fly to Egypt. It was later said that 'Victory in Egypt came by the Takoradi route'.

In the fighting in north-east Africa, RAF headquarters was able to read Italian signals and thus was immediately aware of their intentions. Aided by the South African Air Force, and despite some occasional Italian successes, the RAF provided considerable support for the victory British land forces achieved. The campaign also saw the genesis of improved army-air cooperation.

A Shake Up of Maintenance

In June 1941 the Middle East Air Command came under the command of Air Marshal Sir Arthur Tedder. Also sent to Egypt at that time was Air Vice Marshal Grahame Dawson, an engineer specialist and 'ball of fire', who had been Fighter Command's senior engineering officer during the Battle of Britain. He discovered that hundreds of aircraft were unserviceable because they had not been maintained or repaired.

With Tedder's approval he transformed the salvage and repair operations, including establishing an aero-engine repair depot in the honeycomb of caves near Cairo where the stones for the pyramids had been quarried. As a result, by November 1942 Tedder had over a thousand serviceable aircraft at his disposal compared with a few hundred when he was appointed.

Prior to Tedder's arrival the three armed services had been working with little close liaison. However, he recognised that this was no longer feasible, saying: 'In my opinion, sea, land and air operations in the Middle East Theatre are now so closely inter-related that effective coordination will only be possible if the campaign is considered and controlled as a combined operation in the full sense of that term.' These recommendations were later adopted and made Tedder an ideal choice to become the deputy to General Eisenhower in the later campaigns in the Mediterranean and north-west Europe.

In the Egyptian-Libyan campaigns, when the front line was quickly shifting forwards and backwards, the provision of RAF support to the Army placed enormous strain on its ground support engineers, who had to provide the services needed to maintain and operate the aircraft and airfields. It was said, 'Air forces are no more than idle masses of machinery in the absence of airbases and supplies … The RAF … must learn to be nomadic.'

The RAF (and the Army's) success in becoming so can be summarised by the statistics for the fighting around Benghazi in February 1941. In the air, for a total loss

Dawson, Air Vice Marshal Grahame, CB CBE (1895–1944)

After qualifying as a pilot with the RNAS in 1915 he was appointed technical officer at the Flying Instructors School. In 1919 he was awarded an RAF Short Service Commission in the rank of Flight Lieutenant (Technical) RAF and the following year was appointed technical officer, Engine Repair Depot, Egypt. He was awarded a permanent commission in 1926 and the following year seconded to the Royal Australian Air Force. He was selected to attend the RAF Staff College in 1930 prior to being appointed engineering staff officer, HQ Air Defence of Great Britain. In 1938 he was appointed Command Engineering Officer, HQ Fighter Command, and in 1940 formally transferred to the RAF's Technical Branch. In the Battle of Britain he was Fighter Command's senior Engineering Officer. In June 1941 he was appointed Air Officer (Maintenance), HQ RAF Middle East Command, and became Senior Air Staff Officer, Mediterranean Air Command, in February 1943. He died the following year.

of 26 RAF aircraft, the enemy lost 58 in combat plus some 1,200 captured, while the Army's 31,000-strong force destroyed ten Italian divisions and captured 26 generals, 130,000 troops, 845 guns, and nearly 400 tanks for total casualties of fewer than 2,000.

When General Rommel and the *Afrika Korps* arrived in North Africa in February 1941 to support their Italian allies they were accompanied by *Luftwaffe* aircraft that included day and night fighters, bombers, dive bombers, seaplanes, transports and reconnaissance aircraft. From their bases in Libya they not only provided support for the German and Italian troops on the ground and the Axis convoys sailing between Italy and Libya, but they also attacked Malta and British naval operations.

With increasing numbers of aircraft, air crew and ground support staff arriving, the RAF eventually proved more than a match for the *Luftwaffe*; by doing so, they provided ever greater support for the Army and the Royal Navy, thereby making a major contribution to the eventual victory in North Africa.

The Battle of Matapan – the invisible killers

New technology created another naval precedent at the Battle of Matapan, which took place off the southern Greek coast on the night of 28–29 March 1941, when Italian ships, which had been located in the dark by radar, only became aware of the British fleet's presence when they were being bombarded.

By observing the distance, speed and course of the Italian ships, radar was also able to inform the gunnery officers of the range and direction at which their guns should be set. In the resulting action, the Italians lost three heavy cruisers and two destroyers without any corresponding British loss – 'resulting in the Royal Navy's greatest victory in a fleet encounter since Trafalgar; and, as it was to prove, the last in all its long history'.

DEFENDING MALTA

Crucial to the outcome of the Mediterranean conflict was the Allied defence of Malta.

Maltese Resistance

Submarines and aircraft operating out of the island had inflicted great damage on convoys ferrying vital supplies to Rommel's *Afrika Korps* between south Italian and North African ports. The response of the Axis powers was to try to bomb Malta into submission.

Thus began the siege of the island, whose defence was entirely dependent on the RN and RAF. Furthermore, supplies and reinforcements could only get to the island on convoys sailing from Gibraltar or Alexandria along routes which ran near enemy air bases, were vulnerable to submarine attack and where close air cover could only be provided from carriers. Throughout 1942 Britain endeavoured to concentrate on the defence of the beleaguered island and the fighting through of convoys to take supplies and reinforcements there. Fast ships such as the minelayer HMS *Welshman* sometimes went independently as well.

When Italy entered the war in 1940 there were only ten aircraft based in Malta, of which only four were Sea Gladiator fighters suitable for the defence of the island; one of these was wrecked shortly afterwards leaving only *Faith*, *Hope*, and *Charity* (as the three remaining planes became known). In the following months 14 Hurricanes reached the island having flown off the decks of aircraft carriers although a further eight were lost after becoming airborne.

Despite all the odds Malta survived 1941 even though the *Luftwaffe* had joined the Italian Air Force in attacking the island. Increasingly heavy aerial onslaughts on both the island and the convoys attempting to resupply it continued into 1942.

The RAF's airfields and aircraft were prime targets. In April 47 Spitfires flew off the American carrier USS *Wasp*, no British carrier being available. Their arrival was

Italian Human Torpedoes

In December 1941 the battleships HMSs *Queen Elizabeth* and *Valiant* were seriously damaged by midget submarines while lying in Alexandria harbour and while Admiral Cunningham, the commander in chief, was relaxing on the quarterdeck of his flagship HMS *Queen Elizabeth.*

The two-man crews of Italian human torpedoes were able to attach explosive devices to their hulls. Taken to the vicinity of their target by a mother submarine, two frogmen sat astride a cylindrical battery-driven torpedo. When they arrived underneath their target they detached the nose cone that contained a delayed action mine, which they attached to the ship's hull.

Italian Maiale two-man submarine. http://www.hisutton.com/

however observed by the Germans, and within 20 minutes they were under attack. By next morning only 27 were serviceable – and only 17 by the evening. Towards the end of the month the Governor reported that the island's serviceable fighters had been reduced to six.

The aerial bombardment suffered by the island was comparable to that inflicted on London during the *Blitz*. In April 1942 there were nearly 5,000 raids, with the warning sirens sounding on average every two and a half hours. There was considerable material damage to public and private buildings, but civilian casualties were relatively low thanks to the many underground shelters built into the island's rock foundations as a matter of urgency by over 5,000 miners and labourers after the first major raid in January 1941.

Although the Axis powers considered invading the island this never happened – the losses to the German airborne forces that attacked Crete were so high that Hitler had resolved not to repeat their use.

Relief eventually came through the land victories won by the armies in North Africa, and in the first half of 1943, the island moved from a defensive to an offensive role. It became a major base from which to mount the operations to invade Sicily and the Italian mainland. The bravery and heroism shown by Malta and its population were recognised by the award of the George Cross to the island.

A Ravaged Fleet

Until November 1942 when the siege was lifted, the fighting of convoys through to Malta became a major activity for the Mediterranean Fleet. Although mostly successful, the losses to merchantmen and naval forces were considerable. Indeed, they became so high that convoy operations ceased for a period in the autumn of 1942.

To add to the Mediterranean Fleet's problems, before the end of the year the First World War-vintage battleship HMS *Barham* and a cruiser were torpedoed and sunk by German U boats, while a cruiser and two destroyers were sunk and others damaged by mines off Tripoli. Throughout these early days the naval situation was hampered by the lack of the equivalent of RAF Coastal Command in the Mediterranean.

As well as attacks by surface ships and aircraft, both sides also increasingly used submarines and mines. Successful RN countermeasures against submarines included rocket-propelled projectiles fired from aircraft, which had been developed by DMWD.

Such continual activity not only exhausted the ships' crews but also made it difficult for the naval engineers to find the time for machinery maintenance, boiler cleaning, and other such activities that were essential if the ships were to operate at anywhere near maximum efficiency.

Cleaning the tubes inside a boiler's steam drum. IWM

Exiting the drum. IWM

Engine room of aircraft-carrier HMS Colossus. *IWM*

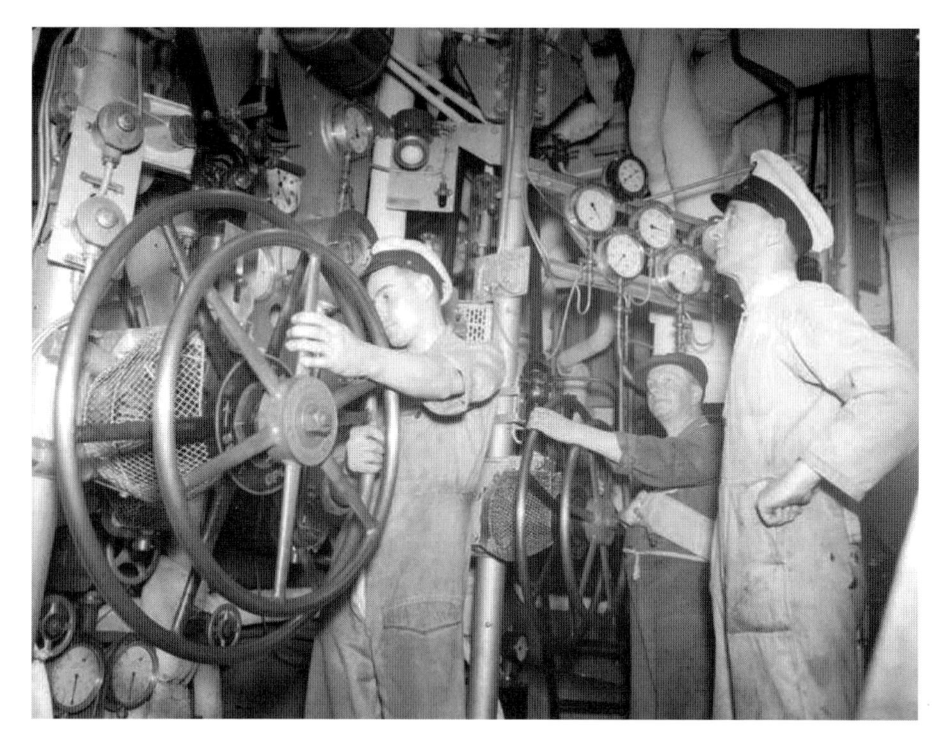

Engine room destroyer HMS Zetland. *IWM*

The Italian Campaign

The invasion of Sicily on 10 July 1943 started the Italian campaign. The operation involved nearly 200,000 British, Canadian, and American troops; some 2,500 ships and major landing craft; over 4,000 aircraft; and vast amounts of artillery, armoured and other vehicles, ammunition and stores, which were assembled from ports in the USA, UK and Mediterranean. Air support for the landings was provided by Allied air forces from bases in Malta and North Africa and by carrier-borne naval aircraft. It took 20 typists seven days to type the originals of the naval operational orders alone, after which 800 copies had to be run off.

Dead Clever

The invasion of Sicily was preceded by the gruesomely named Operation Mincemeat, in which a corpse disguised as a British officer carrying top-secret papers was washed up on the Spanish coast with the object of deceiving the Germans into thinking the landings would be in Greece. The book and film *The Man Who Never Was* describe this operation.

First ashore were men of two commando units. Priority targets for the assault forces included local airfields, from which enemy aircraft could attack the invading forces, and harbours for offloading reinforcements and supplies. The forces ranged against them included some 500 *Luftwaffe* aircraft and the Italian Navy, which still mustered six battleships and seven cruisers as well as destroyers, submarines, and torpedo boats. In the first three days 80,000 men, 7,000 vehicles, 300 tanks, and 900 guns were landed.

The RE's tasks included neutralising beach obstacles and minefields; establishing exits for armour and vehicles onto the local road network; constructing beach maintenance and storage areas; the repair and/or construction of airfields; and generally facilitating the advance of the Army inland. After the landing they were also involved with the operation of the local rail network.

Key to the success of all landing operations were the landing craft in which Army personnel and vehicles could be conveyed ashore. Following a training exercise off Devon in 1938 carried out by the Inter-Service Training and Development Centre, it was realised that existing craft and procedures were woefully inadequate. The landing craft drew 1.5 metres of water, for example, and the exit ramp was far too steep for a vehicle. The Centre then worked out a tactical drill for assault landings which involved building landing ships capable of carrying smaller craft that would take the troops to the beachhead, and specialist vessels capable of carrying and landing armour, transport vehicles, support equipment and stores. From this original conception the multitude of Second World War landing craft evolved.

Building Bridges to Victory

Italy's topography of countless river valleys flowing east and west from the central spine of the Apennines Mountains was ideal terrain in which to mount determined resistance – and engineers played a vital role in the operations to cross them.

Throughout the Italian campaign the RE was engaged in its traditional panoply of services, including: removing explosive charges from bridges whenever possible, locating and neutralising mines and booby traps in roads, bridges and buildings, rehabilitating roads by filling in craters and widening them by the use of bulldozers, bridge building, surveying and map production, providing water supplies and postal services, and adapting and maintaining barracks. Many other RE personnel were engaged in restoring and operating the railway network, while others supervised the management of Italian factories that were taken over and used to manufacture products for the Allied forces. Following the Taranto landing the RE also repaired the Apulian Aqueduct which conveyed water across the Apennines from the west to the dry 'heel' of Italy.

The major task of the RE and the US Corps of Engineers, however, was the construction of the thousands of assault bridges needed to enable the armies to advance

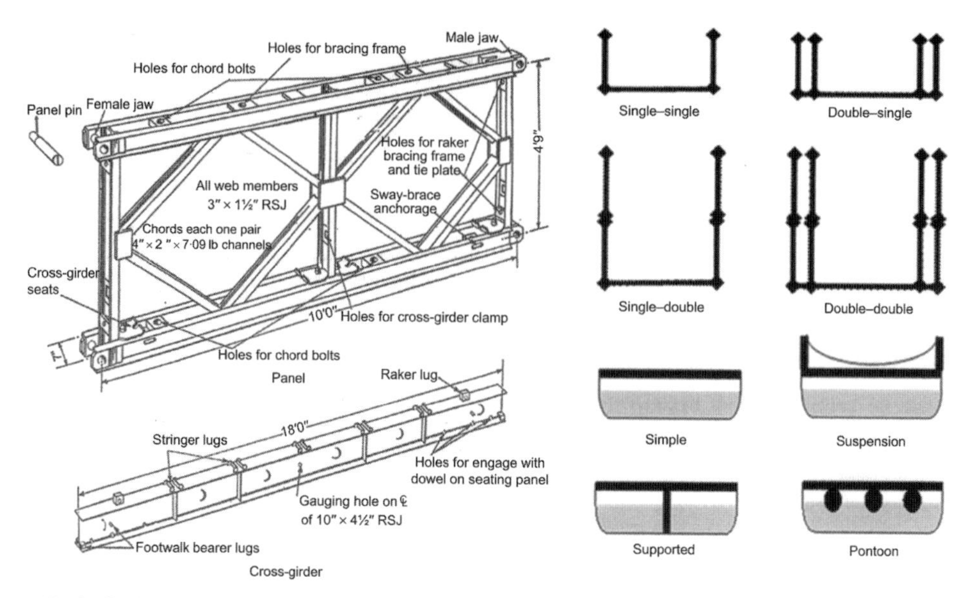

Bailey bridge: panel, possible modes, and configurations (1' = 305 mm, 1" = 25 mm). IWM

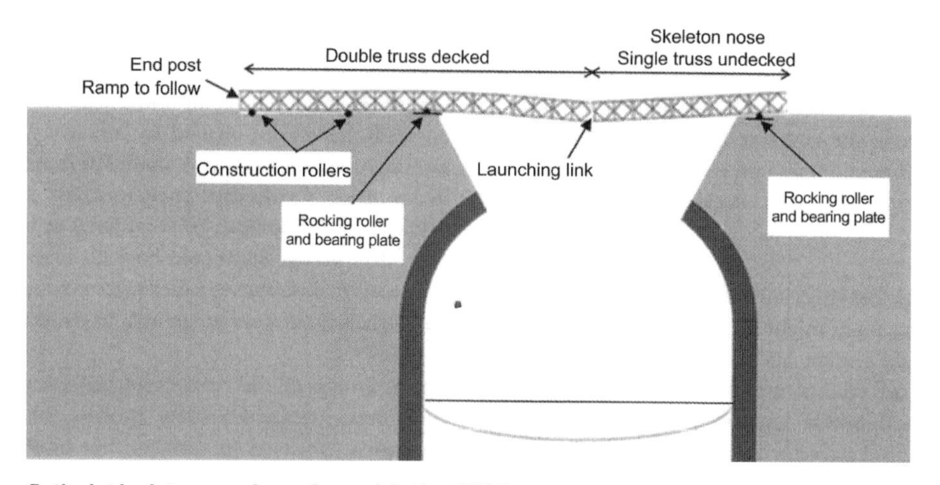

Bailey bridge being erected to replace arch bridge. IWM

across the river valleys that confronted them. For this the versatile Bailey bridge proved to be the ideal instrument – so much so that it was used by US Army engineers in preference to their own.

The Bailey bridge had been adopted as the British Army's main bridging resource in 1941, and it was first used in Tunisia the following year. Conceived and developed by Donald (later Sir Donald) Bailey, a civil engineer employed at the RE Experimental Bridging Establishment in Dorset, it comprised 3-metre-long steel panels capable of

Bailey, Sir Donald OBE FICE (1901–85)

A civil engineer, he designed the bridge that bears his name. Field-Marshal Montgomery is recorded as saying that 'without the Bailey bridge, we should not have won the war.'

After schooling in Rotherham and Cambridge, Bailey studied at the University of Sheffiled and was employed as a civilian at the War Office's Experimental Bridging Establishment at Christchurch, Dorset. The bridge he created was capable of being rapidly assembled in a variety of modes and configurations; none of the bridge components was heavier than a 6-man lift. By the end of the war, over 4,000 bridges totalling some 350 km of fixed and 60 km of floating pontoon Bailey bridges had been built by engineers of the British and Commonwealth armies in North Africa, Italy, north-west Europe, and the Far East. The US Army also adopted it in preference to their own comparable design and as an alternative to their own timber bridges. In addition to the UK, manufacture of the components took place in Canada and India.

General Eisenhower said that 'Along with radar and the heavy bomber, it was one of the three most important engineering and technological developments of WW2', while Field-Marshal Montgomery wrote: 'Bailey bridging made an immense contribution towards final victory in World War II. As far as my own operations were concerned with the Eighth Army in Italy and with 21 Army Group in N. W. Europe, I could never have maintained the speed and tempo of forward movement without large supplies of Bailey Bridging.'

Various versions of the bridge have been assembled for civilian use in various parts of the world since the war. Bailey was knighted in 1946 and awarded £12,000 by the Royal Commission on Awards to Inventors. Post-war he served on the council of the Institution of Civil Engineers.

being handled by six men and of being rapidly erected in a variety of configurations. In addition to being suitable for spanning short distances as a simple bridge, it could also be used for longer crossings, either with an intermediate support or supports on floating pontoons. It was also occasionally used as a suspension bridge.

When a bridge was considered necessary the local military commander and senior RE officer would determine its location and class i.e. how much weight it could carry. It was necessary for a small advance party to cross the river by boat or by wading in order to secure a beachhead on the far bank. The first priorities of the RE officer assigned to take charge of the bridge's construction were to determine the equipment needed and the order in which it should arrive on site, and then to prepare the assembly, launch, and landing areas on the two banks. The bridge would then be assembled prior to being launched as a cantilever.

Light conditions were a major consideration when bridge building; working in total darkness without being able to use torches was virtually impossible. A clear moonlit night would enable the complex work to be carried out within the timeframe. This was normally dusk to dawn or earlier, allowing armour and support vehicles to cross and reinforce the bridgehead before the sun came up. Sometimes artificial moonlight could be provided by reflecting searchlight beams off the clouds. The construction was hard manual work, and it always took place as quietly as possible without mechanical means.

A reserve construction unit stood by, under cover some way back, to continue with the erection in the event of heavy casualties. Spare equipment was also left on site to repair any later damage. None of this could be done without the skill, determination and bravery of well-trained sappers. As was said to a general seeing sappers working under shellfire: 'You can't build a bridge lying down, sir!'

A 24 metre bridge over the Rapido River; the bridge was completed in 12 hours during the night of 12/13 May 1944, at a cost of 15 killed and 60 wounded. Painting by Terence Cuneo (by permission of the Institution of Royal Engineers)

Despite the difficulties, it was usually possible to complete a bridge in a single night (and later statistics from operations in north-west Europe showed that, providing there was sufficient labour, the average time to construct and set up a standard Class 40 bridge was one hour per 3 metres of length).

Bridge construction troops enjoyed considerable satisfaction and relief when a successful touchdown on the far bank's rollers was achieved – the first vehicles had crossed safely and casualties could be brought back from the bridgehead.

A Bailey pontoon bridge was built across the River Po, but it was later decided to replace this with a semi-permanent bridge. Several alternative proposals were considered, but at that time there were misgivings about using Bailey units for long-span continuous bridges. After detailed calculations had been carried out and a trial bridge built and tested, the four-span, 300-metre-long treble-single Springbok Bridge was built by South African engineers and completed in six days.

Working behind the front line, railway construction companies helped to re-establish the Italian rail network which required the construction or repair of many bridges. Although Bailey equipment was used initially, later construction mostly used specially designed or captured German equipment. These operations continued well after the fighting had ended, as did the improvements carried out to the road network.

Including the semi-permanent road and rail bridges built behind the front line, some 2,800 Bailey bridges, totalling around 70 km in length, were constructed in Italy. All but 19 were simple 'dry' bridges, as pontoon bridges were needed only for some of the wider crossings.

Signalling Duties

The Italian campaign was the first major campaign in which the Royal Signals had to work under Allied command headquarters and in close conjunction with their American counterparts. There were also several novel circumstances, such as combined service operations, amphibious and airborne landings, and the need for close liaison between the Army and tactical air force units. The official Corps history summarised its achievements thus:

> *The main types of operations with all their attendant factors produced a great variety of situations. Aided by the invaluable experience of the earlier Middle East campaigns and the large proportion of seasoned signal units, the theatre can justly claim to have evolved a communication system combining a sound balance of the various means with the maximum economy of resources.*

Standardisation of signal practice with the Americans was achieved, although there were difficulties arising from the use of separate codes and cyphers. Submarine cables were modified so as to carry press reports and the troops' personal telegrams. The fighting seriously damaged the Italian overhead telephone networks and, while underground cables were less affected, repeater stations had generally been sabotaged before the Germans retreated, often by being set alight after being doused in petrol; mobile repeater stations were frequently needed.

Radio links were established for the crossing of the Straits of Messina, and wireless communications were used in these operations. Due to the high volume of signal traffic, operating schedules were established and great pains taken to ensure that variations in volume did not disclose information regarding future operations. Teletype was increasingly used, and a radio teletype company was formed. Unlike the Western Desert, the Italian terrain precluded large-scale battles between armoured units, yet communications links were required between tank and infantry units when the former were providing close support to the infantry. Many links were required between the headquarters and units of the tactical air forces and armies.

REME Activities

The preparations for the invasion of Sicily involved the establishment of Beach Maintenance Groups, the operation of which was rehearsed in great secrecy in the Gulf of Aqaba some weeks before the landings. The REME component was originally intended to recover and repair vehicles drowned in the amphibious landings, although its role evolved into maintaining a speedy throughput of equipment across the beach. Thereafter its main task was the repair and recovery of vehicles and artillery as it had been in North Africa; it was also responsible for modifying tanks to make them suitable for the Italian terrain.

Prior to the landings in Sicily it was appreciated that waterproofing vehicles would be necessary, and REME headquarters in the UK evolved a five-stage programme which included work at the assembly ports. In the event over 15,000 vehicles were waterproofed at various bases around the Mediterranean.

Although the British Eighth Army was mainly engaged on the Adriatic side of the Apennines, it was decided to establish REME's main engine overhaul centre at Naples. This had an estimated output of a thousand engines a month; it was staffed by some 2,000 military and up to 8,000 local civilian personnel.

Base manufacturing workshops were established to provide essential spares and modification kits as well as special items. They removed guns from tanks and converted them into troop carriers or assault and bridging vehicles, and made 12,000 'grousers' for fitting to tank tracks operating in muddy conditions (and 900 artificial eyes for use by the medics prior to the later fitting of a permanent glass eye!). Various advanced workshops were established near the front line. These were not large but were very versatile and adaptable, and they could be operational within four days of being established.

As with other operations in Italy, the weather caused major problems. For instance, due to excessive rainfall vehicles in a recovery park sank axle-deep in the mud prior to becoming icebound due to freezing weather.

Before the end of the Italian campaign the Western allies landed in Normandy, where the hard lessons learnt in North Africa and Italy would be tested in the struggle towards victory in north-west Europe...

NORTH-WEST EUROPE TIMELINE

1942

26 Jan	US troops start arriving in UK.

1944

May–3 Jun	Troops, vessels, and equipment assigned for Normandy landings moved to launch sites. All troops 'confined to barracks'.
4 Jun	Adverse weather forecast causes invasion to be postponed by 24 hours (from 5th to 6th Jun).
5 Jun	Invasion fleet and airborne assault troops leave bases.
6 Jun	D-Day. Key bridges in beachhead captured or destroyed by airborne forces. Seaborne landings at five beaches in Normandy by UK, Canadian, and US invasion troops.
13 Jun	First V1 flying bomb lands in UK.
26 Jun	Cherbourg captured (but only after all dock installations destroyed by Germans).
9 Jul	Caen captured (following stiff German resistance and Allied aerial bombardment).
20 Jul	Failed assassination attempt on Hitler.
12 Aug	Germans start retreat from Normandy.
15 Aug	Allies land in Southern France.
23 Aug	Paris liberated; Marseilles captured.
2 Sep	Allies enter Belgium.
3 Sep	Brussels and Lyons liberated. Last V1 bomb launched.
4 Sep	Antwerp captured.
8 Sep	First V2 rocket bomb lands on London.
11 Sep	Allies cross Dutch and German frontiers.
17–26 Sep	Abortive Allied attempt to capture key bridges at Arnhem.
1 Nov	Allies land on Walcheren Island to clear channel to Antwerp.
26 Nov	Antwerp port useable.
16 Dec	Germans counter-attack in Ardennes region of Belgium.

1945

25 Jan	Germans defeated in Ardennes.
5 Mar	Cologne captured.
7 Mar	Bridge over Rhine at Remagen captured intact.
27 Mar	Last V2 rocket bomb lands in England.
12 Apr	President Roosevelt dies, succeeded by Truman.
30 Apr	Hitler commits suicide in Berlin bunker.
1 May	Admiral Doenitz becomes German Head of State.
2 May	Berlin surrenders to Russians.
7 May	Unconditional surrender by Germany.
8 May	VE Day.
12 Jul	UK and US troops assume control of two sectors of Berlin (Russia and France in charge of the other two).

Chapter 9

THE BEGINNING OF THE END

D-Day and beyond

The Normandy landings of 6 June 1944 which prepared the way for the eventual Allied victory in north-west Europe were enabled by a perfect fusion of physical courage and engineering ingenuity. The lessons learned from earlier raids and campaigns had inspired the development of new equipment and fresh concepts of warfare that not only empowered the Anglo-American-Canadian invasion but also helped reduce casualties.

The early preparations for D-Day had been undertaken by staff working under Lieutenant-General Frederick Morgan (a Briton who was chief of staff designate to US General Dwight Eisenhower, the Supreme Allied Commander in Europe) and Admiral Lord Louis Mountbatten, the chief of Combined Operations. They came up with many novel and imaginative solutions to the multitude of seemingly intractable problems involved in breaching Hitler's much vaunted 'Atlantic Wall'. Morgan and Mountbatten benefited from their combined experiences, and they also knew that the invading forces would have unstinted resources of personnel and equipment, much of which would be purpose-built. Early consideration was given to the Calais area, but the area of Normandy between Cherbourg and Caen was finally adopted. See map on p. 126.

The landings themselves were supported by many original devices. Some 24,000 parachutists and glider-borne troops, conveyed by nearly a hundred aircraft squadrons and hundreds of gliders, landed shortly after midnight. The seaborne invasion force comprised over 150,000 troops carried in hundreds of tank- and infantry-landing craft which were protected by over 500 escort vessels and minesweepers.

Some 1,500 more ships bombarded the German defensive positions as well as providing additional protection and support for the seaborne force. Allied air forces flew around 140,000 sorties to bomb the German defences, and they also mounted diversionary raids. The Allies suffered some 10,000 casualties on the first day, including over 4,000 killed, while the Germans, fighting from their prepared defensive positions, suffered about 1,000 casualties. Over the next four days over 300,000 troops, 54,000 vehicles, and 100,000 tons of supplies were landed on the beaches.

Operation Tarbrush

On the nights of 18–19 May, Lieutenant F. M. Berncastle DSO RN had surveyed the Arromanches beach from a small boat and taken soundings which provided vital information for the planners of the invasion as well as the Mulberry harbour designers. DMWD designed beach gradient meters for use in these reconnaissance missions.

Aerial reconnaissance showed that unknown objects had been placed on top of anti-invasion posts erected on the Normandy beaches by the Germans. In Operation Tarbrush small detachments were sent to investigate, being transported to the beaches by submarine and small boat. Two of the raiders – Lieutenant Roy Wooldridge RE and the Hungarian-born Dyuri Lányi, now a commando and going under the name of Lieutenant (later Colonel) George Lane – were discovered, taken prisoner, and interviewed by General Rommel, who was in charge in Normandy. To hide his Hungarian accent Lane pretended to be Welsh. Contrary to Hitler's order, they were not executed (and both were later awarded the MC).

Among the officers who returned safely to the UK was Lieutenant John Stone RE, who reported that the objects were mines. As the mines could hole landing craft, the military commanders decided to change the date and time of the D-Day invasions so that the craft reached shallow water seaward of the posts, even though this would involve a longer crossing of the beach. Stone was awarded the MC.

Stone, Lieutenant John MC RE (1922–2011)

He led one of four parties of Royal Engineer volunteers sent shortly before D-Day to reconnoitre the Normandy landing beaches. General Montgomery's original plan was to land at high tide so as to minimise the distance troops would have to advance over open beaches. Aerial reconnaissance, however, disclosed that among the obstacles the Germans had constructed on the beaches were poles on which they had placed objects that were believed to be mines. As these would be underwater at high tide they could hole the bottoms of landing craft. Of the four reconnaissance parties, two were captured and one could not land, but Stone's party got aground, surveyed the shore, and examined the objects and confirmed that they were plate-shaped mines. He was awarded the Military Cross by General Montgomery. Post-war he became a partner in the consulting engineering firm of Mouchel.

Stagg, Group Captain James CB OBE FRSE RAFVR (1900–75)

A Scot, he graduated from the University of Edinburgh and then became an assistant in the Meteorological Office. He led the British Polar Year Expedition to the Canadian Arctic in 1932–33 and was appointed superintendent of the Kew Gardens Observatory in 1939. He joined the RAF Volunteer Reserve in 1943 with the rank of group captain. Post-war he served as director of services at the Meteorological Office until 1960. He was elected as a fellow of the Royal Society of Edinburgh in 1951 and president of the Royal Meteorological Society in 1959.

Set Fair

The successful execution of the landings required a full moon to illuminate obstacles and landing places for gliders; a low tide at dawn to expose the elaborate underwater defences; and acceptable weather conditions. High winds and rough seas could capsize landing craft, thick cloud cover could obscure the landing and drop zones and make the air support difficult, while wet weather could bog down the army. Such optimum combination of moonlight, tide and weather occurred but rarely, and D-Day was originally set for 5 June, the first date in a narrow three-day window expected to have the necessary conditions.

Although dawn on 4 June broke bright and clear and with light winds in Portsmouth, where the Allied commanders were based, the chief meteorologist, Group Captain James Stagg RAF, believed stormy weather would arrive within 24 hours. As well as reports from numerous weather stations, Stagg had several meteorologists advising him, but they were in disagreement. Some, having studied historic weather maps, believed 5 June would be satisfactory, while others, basing their observations on reports from weather stations in the Atlantic, including some from as far afield as Newfoundland, advised otherwise.

Doubtless aided by his prior experience of conditions in the Channel, Stagg concluded that the weather would be unacceptable on the chosen date, although he believed that it would have improved by the following day. He therefore recommended to General Eisenhower that the invasion be postponed by 24 hours. His advice was accepted, and early on 4 June the decision was taken that 6 June should become D-Day.

An unexpected bonus of the postponement was that the German forecasts for both days were for such bad weather that General Rommel decided there was no possibility of an invasion and went to Germany to celebrate his wife's birthday; his deputies were extraordinarily slow in mobilising the defences in his absence.

Had the Allied meteorologists got it wrong, and had the invasion failed because of the weather, the war in the West could well have been lost. No wonder Stagg's prediction has been called 'the most important weather forecast in history'.

Deception and Camouflage

Operation Fortitude was an extremely wide-ranging – and highly successful – plan to deceive the enemy, firstly by concealing the date and location of the Normandy attacks, and secondly, by tricking the Germans into thinking that these were a diversion and that the main invasion would follow across the Strait of Dover to the Pas-de-Calais area. Bletchley Park decrypts informed the Allies that it was two weeks after D-Day before Hitler and the German High Command realised their mistake and at last authorised the movement of army divisions from the Calais area to Normandy.

The deception necessitated the use of a number of subterfuges, including double agents to transmit messages giving false information (made possible because every German agent sent to Britain during the war had been captured shortly after arrival and the majority of them had been turned to become double agents). One double agent, a Spaniard Juan Pujol Garcia, codenamed 'Garbo', was actually awarded an MBE by Britain and the Iron Cross by Hitler!

The transmission of recorded radio traffic and the building and placing of dummy models of landing craft, tanks and guns in Scotland and South East England to suggest that there were invasion forces in these locations poised to attack Norway and the Calais area were also used. In the British Army camouflage had always been the RE's responsibility. The dummy artefacts were mostly made of rubber sheeting draped over aluminium frames. In addition, on the night before D-Day the RAF dropped 'window' or 'chaff' (strips of aluminium) which reflected radar signals, again to give the impression that large numbers of bombers and/or airborne troops were crossing the Strait of Dover. The window dropped by the RAF was the brainchild of Dr Reginald Jones, the Air Ministry scientist who had 'bent the radio beams' during the *Blitz* following the Battle of Britain (see Chapter 4).

Operation Fortitude undoubtedly made a major contribution towards the success of the Normandy landings and has been described as 'the greatest orchestration of deception ever achieved in military history'. Churchill said, 'In war-time, truth is so precious that she should always be attended by a bodyguard of lies.' The success of Operation Fortitude bears this out to a remarkable degree.

Indeed, so effective were these operations that the Germans actually increased their forces in Norway in the weeks preceding D-Day. They also believed that the Allies had 90 divisions in Britain ready to invade the Continent, when in reality there were only 37. To a degree, the false information planted by the double agents served to reinforce German preconceptions about the invasion. When planning Operation Sea Lion for the invasion of Britain in 1940 their main intended thrust was across the Strait of Dover, and when considering how the Allies would invade France in 1944, they thought that they would most probably choose the same route. In part, also, this was a failure to appreciate the flexibility afforded by control of the sea.

Bridgemanship

One of the earliest actions of D-Day was the successful capture by a glider-borne assault force of two bridges over the Caen Canal and River Orne. This enabled parachute troops landing later that night to link up with the seaborne invasion force and thereby protect its flank from counter attack. The bridges were subsequently named Pegasus and Horsa in honour of the airborne troops involved. Two new bridges were also built over these waterways by RE units of the invasion force, the first being completed within 60 hours of the first arrival and the second two days later.

In the early hours of D-Day, Major Tim Roseveare RE parachuted into Normandy in command of an RE squadron with orders to destroy four bridges on the beachhead's eastern flank. Despite coming down around midnight some 8 km away from their intended dropping point, and having to fight a gun battle with the Germans in which they sustained heavy casualties, the squadron achieved all its objectives. They even met a French farmer and told him that he had been liberated; he was unimpressed and carried on milking his cow! The survivors arrived at brigade headquarters as a very bedraggled and exhausted party, having been shot at by the Germans, bombed by the RAF, shelled by the Navy and unappreciated by the French.

Roseveare was awarded an immediate Distinguished Service Order, and three Military Crosses and two Military Medals were awarded to members of his squadron. The squadron subsequently built an assault bridge where they had previously destroyed

Pegasus bridge over the Caen Canal (subsequently replaced by a wider replica) (by permission of the Institution of Royal Engineers)

Bridges over the River Orne (by permission of the Institution of Royal Engineers)

Roseveare, Major J. C. A. (Tim) DSO RE (1914–2000)

He was awarded an immediate DSO for his leadership and achievements after parachuting into Normandy in the early hours of D-Day in command of a squadron. Their instructions were to destroy four bridges that the Germans might use to reinforce the troops resisting the invasion landings. He landed some 8 km from his intended drop zone; after collecting those members of his squadron he could find and commandeering a Jeep, they fought and won a gun battle with German forces, and then managed to achieve all their objectives. Their duties completed, Roseveare ditched the Jeep in a nearby flooded field and swam through several water courses before taking to the woods. 'Eventually,' he recalled, 'we came across an elderly Frenchman milking a cow. When I told him he had been liberated, he was not impressed. Perhaps he could not understand my accent.' The remnants of his squadron finally arrived at Brigade HQ at midday.

Roseveare read engineering at King's College London before working for a small civil engineering consultancy. On the outbreak of war he was commissioned into the Royal Engineers and went to France with the British Expeditionary Force and was nearly taken prisoner before returning to Britain. After D-Day his reformed squadron built an assault bridge to replace one of those they had previously destroyed. They subsequently participated in other actions, including the Battle of the Bulge in the Ardennes at Christmas 1944, before advancing eastward and finally meeting up with the Russians just before VE Day. Post-war he worked on supply and hydroelectric projects in both the UK and overseas for consulting engineers, firstly for Binnie & Partners and then Freeman Fox and Partners (with whom he later became a partner). He frequently returned to Normandy for D-Day anniversary celebrations and one of the streets in Troam has been named after him.

Light as Air

American gliders were mostly WACOs while the British mainly used Horsas or Hamilcars, which could carry 13, 29, or 40 fully kitted troops respectively or an equivalent weight of vehicles or weaponry.

one, and went on to see action in a number of operations, finally crossing the Rhine and meeting up with the Russians on the Baltic coast just before VE Day.

Operation Neptune

The naval operations, codenamed Operation Neptune, were under the overall command of Admiral Sir Bernard Ramsay, who had masterminded the Dunkirk evacuation and planned and organised the landings in North Africa and Sicily. The main objective of the naval forces was the safe transport of the seaborne invasion troops and equipment. Additionally, they bombarded German defensive positions, attacked any German naval activity threatening to interfere with the attack, and ensured the safe and effective deployment of the vessels involved in the Mulberry harbour and PLUTO operations.

The planning of the movement of five divisions of troops and over 4,000 vessels in three days was a mammoth undertaking; the operational orders comprised 22 parts and ran to 579 pages plus numerous appendices. The introduction stated: 'This is probably the largest and most complicated operation ever undertaken.'

Ramsay recognised that the Navy's task was to deliver troops and stores as, when, and where the Army required them. To that end he and General Montgomery, with whom he had collaborated in the preparations for the Sicily landings, held frequent meetings. Among Ramsay's major concerns were the minefields laid by the Germans and the need to ensure that ships arrived and disembarked in the sequence required by the Army.

Although by 1944 the Royal Navy had expanded from 127,000 officers and ratings in 1939 to some 750,000 men plus 55,000 Wrens, the demands of the Normandy operation were expected to require a further 35,000 sailors and 10,000 Wrens. Part of the heightened demand was because of the greater numbers of electrical and radar personnel required to operate the increasingly sophisticated equipment now carried by warships. To solve this problem, a number of would-be soldiers and airmen found themselves in the naval service.

From around 45 days prior to D-Day, both the Navy and RAF mined Continental ports from the Baltic Sea to the Bay of Biscay. Also, for many months before D-Day the Allied naval, land and air forces rehearsed the embarkation, transit and landing of troops around the British coast. While highly beneficial and mostly successful, one embarkation and transit exercise of American troops from Slapton Sands, Devon, in April 1944 went disastrously wrong, with German gunboats causing many casualties due to a breakdown in communications between the Americans and the RN force assigned to protect them.

For the actual invasion, five naval units, each carrying one of the five assault forces, assembled in southern English ports; when ordered to sail they went first to the assembly point 'Z' south of the Isle of Wight before travelling along 'the spout' to Normandy. Preceding these forces were the minesweepers which cleared, and marked

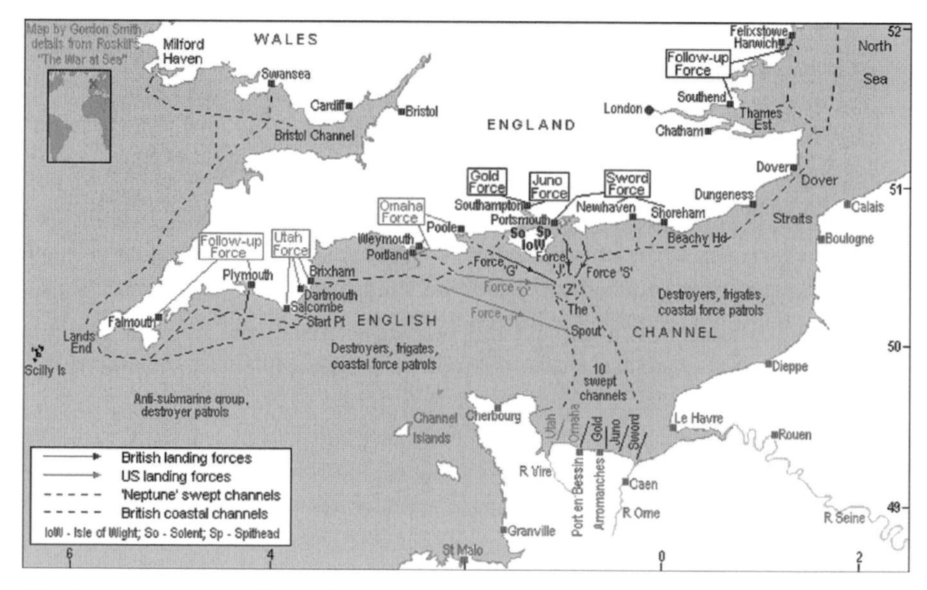

Naval operations plan for Normandy invasion. Map from Gordon Smith, details from Roskill's The War at Sea

with flashing buoys, ten channels through the German minefield – five for southbound vessels and five for vessels returning to the UK. Anti-submarine and coastal forces support groups and two reserve follow-up forces were also established. Each of the five assault forces had its own headquarters ship, which, bristling with communications paraphernalia, commanded their naval, air force and army operations until onshore facilities could be provided.

The invasion fleet carried 130,000 troops, 2,000 tanks, 12,000 vehicles, and nearly 10,000 tons of stores and was manned by almost 200,000 RN and MN sailors. Ahead of these forces were two midget submarines; having waited offshore for three days mostly lying on the seabed, they surfaced at 5 a.m. on D-Day three miles from the beaches and started flashing green lights seaward so as to guide the invasion forces.

So confident was the Allied High Command that they would have complete domination of the sea and skies that all vehicles and aircraft had large readily visible white markings painted on them just prior to D-Day. That confidence was justified as neither the German Navy nor the *Luftwaffe* was able to seriously interfere with the campaign, and Allied losses of vessels and aircraft were minimal. Only one minesweeper was sunk and two vessels damaged by conventional mines.

The sweeping of conventional mines outside 'the spout' continued over the following three months, and it has been estimated that the number disposed of during this period totalled 10% of all the mines swept during the previous five years.

However, the new 'oyster' (or pressure-operated) mines dropped at night by the *Luftwaffe* initially caused more serious damage, with seven destroyers or smaller warships being sunk and a number more damaged. Fortunately, after a means of combating magnetic mines had been developed in 1940, Admiralty scientists had investigated and developed solutions to other potential non-contact mines (see Chapter 3). As a result they were aware that these mines were activated by changes in the water pressure resulting from vessels passing over them, and that the best countermeasure was for vessels to slow down in shallow water. After the Admiralty issued instructions to this effect, losses decreased dramatically. It was also found that changes in water pressure caused by tidal swells often detonated them harmlessly.

That the invasion was a complete technical and strategic surprise to the Germans was in part due to the electronic jamming of German radar stations. Although the airborne landings and aerial bombardment had begun around midnight, it was not until 3 a.m. that the Germans were aware that there was an armada at sea approaching Normandy.

Mulberry Harbours

The conception and planning of the two artificial Mulberry harbours had begun shortly after the disastrous Dieppe raid of August 1942, which had shown the impossibility of capturing a working port and of maintaining supply lines across open beaches. In a famous memo of 30 May 1942, Churchill wrote:

> *PIERS FOR USE ON BEACHES. They must float up and down with the tide. The problem must be mastered. Let me have the best solution worked out. Don't argue the matter. The difficulties will argue for themselves. WSC.*

In March 1943, the Combined Chiefs of Staff sent a memo to the First Sea Lord:

> *This project [artificial harbours] is so vital to ... [the invasion operation] that it might be described as the crux of the whole operation.*

At the Quebec conference of August 1943 the Anglo-American leaders approved proposals submitted by Brigadier Bruce White, the War Office Director of Ports, for the construction of two of these harbours.

Some of Britain's leading civil engineers were given temporary naval or military ranks while developing the detailed designs under the overall direction of the Admiralty and War Office. Among them was John Jellett, an Admiralty civil engineer who had been involved with the design or construction of many of the RN's support facilities worldwide. Working with only pre-war charts supplemented by the results of clandestine reconnaissance missions carried out by RN and RE parties, and with extremely tight deadlines to meet, they had to produce reliable solutions to unusual and complex problems. The National Physical Laboratory and other research establishments undertook experiments for the planners. Prototypes of some elements were also built

White, Brigadier Sir Bruce, KBE FCGI FICE FIEE (1885–1983)

One of the leading British consulting engineers of his generation, he masterminded the design and construction of the two Mulberry harbours in the Second World War. He served with the Royal Engineers in the First World War and was involved in the design and construction of Richborough military port near Sandwich, Kent. The son of consulting engineer Robert White, he joined his father's firm in 1919 and became senior partner on his father's death. He later merged the firm with another consultantcy. He was knighted for his wartime service and post-war the practice became known as Sir Bruce White, Woolfe Barry and Partners. They designed the Chiswick flyover, Bombay Marine Oil Terminal, Damman Port, Muara Port, the UK's first container terminal at Tilbury, and Singapore's first container berth. The firm later became the marine consulting arm of the Acer group. He died on 29 September 1983 having worked into his nineties.

Tennant, Admiral Sir William KCB CBE MVO DL (1890–1963)

He entered the Royal Navy aged 15. In the First World War he saw service in Gallipoli and at the Battle of Jutland when his ship was one of those that were sunk. He served in the battlecruiser HMS *Renown* for the Prince of Wales' overseas tours in the 1920s before being promoted to captain in 1932. Working at the Admiralty during the fall of France he hurriedly made plans for the Dunkirk evacuation before being sent to the beaches to organise the naval side of the operation. As the last person to leave Dunkirk, he scoured the beaches with a megaphone searching for any troops still on shore. Awarded the CB for his efforts he also earned the nickname of 'Dunkirk Joe'. Appointed to command his old ship the HMS *Renown* he saw service in the Atlantic before the ship was sent to the Far East following the Japanese attack on Malaya. However, on arrival there, both the *Renown* and HMS *Prince of Wales* were attacked by Japanese aircraft and sunk. Back in the UK he was promoted to admiral and appointed to supervise the naval side of the transport, assembly, and set up of the two Mulberry harbours, and the laying of the cross-Channel PLUTO submarine pipelines. Post-war he served as commander in chief of the Royal Navy's America and West Indies Squadron.

Jellett, Dr John FICE OBE (1905–71)

Having graduated from the University of Cambridge he first worked with civil engineering consultants and was involved in the designs for bridges in the UK, India, and Fiji and in upgrading a major British canal. In 1933 he joined the Admiralty's civil engineering department. Two years later he was appointed civil engineer at Singapore Naval Base where he was in charge of major works to provide new facilities there. In 1938 he returned to the UK. After being occupied with major works for Fleet Air Arm bases and graving docks at Devonport and Gibraltar, he was appointed officer-in-charge of the Admiralty operations at Milford Haven, Pembrokeshire, prior to being involved with sea forts and various onshore naval projects. He was then employed as superintendent civil engineer in the Eastern Mediterranean where he was responsible for projects in Egypt and the Levant. His next posting was to Malta to restore the dockyard following the damage caused by bombing during the siege. He then returned to the UK to take charge of the construction and assembly of the concrete caissons that formed part of the Mulberry harbour breakwaters. He retired from Admiralty service in 1946 and moved to Southampton as a docks engineer, becoming chief docks engineer in 1958. In this post he oversaw the repair of war damage, the reclamation of land from salt marshes, and the diversion of the River Test so as to build 30 additional berths. Retiring in 1965 he became a consultant to Giffords and served as president of the Institution of Civil Engineers in 1968–69.

and tested. The DMWD considered various solutions that might assist the operation as well. The technical considerations and the solutions developed were described in the last seven papers published in volume two of the Proceedings of the Institution of Civil Engineers' 1948 conference on 'The Civil Engineer in War'.

Each of the two Mulberry artificial harbours enclosed an area measuring some 3 km by 1.5 km and was thus about the size of Dover harbour. The structural components were constructed at coastal locations around Britain and towed across the Channel prior to assembly off Normandy. Each harbour consisted of breakwaters and floating pierheads and roadways to connect them to the shore. Floating Bombardon breakwaters supplemented the protection provided in the critical first few days while the harbours were being assembled. Admiral William Tennant oversaw the naval side of the transport, assembly and set-up of the two harbours.

The three main elements of the British Mulberry B harbour were: breakwaters (containing rectangular concrete caissons and sunken blockships); floating spud

Mulberry Harbour B at Arromanches after assembly. ICE

pontoon pierheads (where landing ship tanks (LSTs) could discharge vehicles, ships could offload stores, and barges conveying stores from anchored vessels could discharge); and floating roadways or 'whales' (to connect the pierheads with the shore). The American Mulberry A harbour was similar, except that there were small differences to the pierheads and breakwaters.

Assembly off Normandy began the day after D-Day and the first troops and stores were landed five days later. Two weeks after D-Day the severest storm for 40 years hit the Normandy coast. Unfortunately the American harbour was destroyed but the British harbour, although damaged, survived and became the main supply route for both the British and American armies until the port of Antwerp could be liberated and made operational in mid-November. Over the months it was in use, Mulberry B (known as Port Winston) landed 2.5 million troops, 500,000 vehicles, and 4 million tons of supplies.

The harbours were located where they would not interfere with the military aspects of the initial landings and where transport leaving the harbours could access the local road network. The rectangular concrete caissons were used in the shallower waters and the blockships in deeper locations. Most of the caissons were 62 metres long with heights varying between 7.5 and 18 metres. The tidal range at the site was 8 metres and the pierheads had to be located where there was a minimum depth of water of 5.5 metres, while anchorages with a minimum depth of 7 metres were required inside the breakwaters. The caissons were towed across the Channel but the blockships sailed across independently. Anti-aircraft defences were mounted on the caissons.

The pierheads comprised floating pontoons with holes in the corners through which slotted spuds, or legs, were lowered to the seabed, thus enabling the pierheads

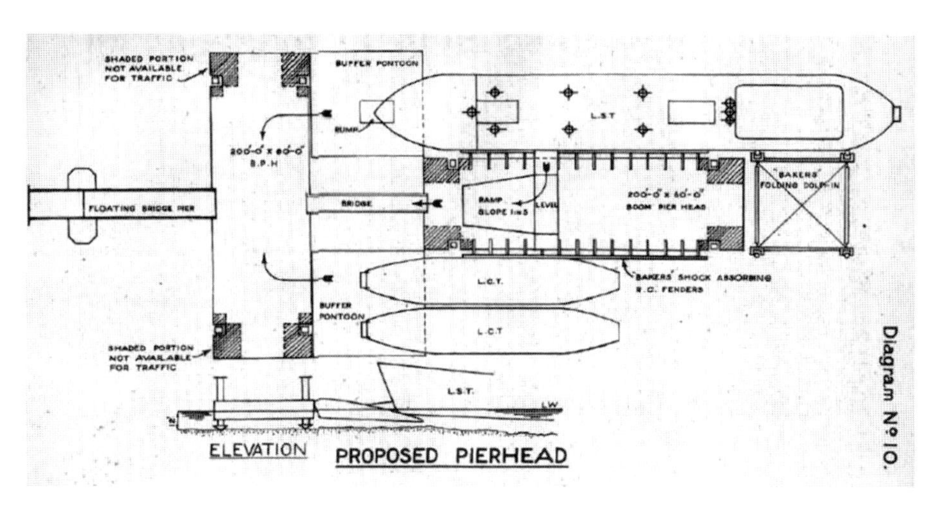

T-shaped pierhead comprising two pontoons for offloading LSTs. ICE

LST pontoon with reinforced concrete fender and platform with ramp raised. ICE

Ambulance exiting bows of LST on to buffer pontoon. ICE

to rise and fall with the tide. Two types were constructed; one for taking stores from ships berthed alongside and another at which armour and vehicles could land from LSTs. Two LSTs could berth end-on at these, and unique reinforced concrete fenders and ramps were incorporated so that vehicles could land from the upper as well as the lower decks.

The designs had to account for each pontoon being self-contained with its own power plant as well as being seaworthy, easily towable, and capable of operation with tidal currents and waves 1.5 metres high. Ships up to 90 metres long and with draughts of up to 9 metres were to be accommodated. With a tidal range of 8 metres the pierheads had to be effective in water up to 15 metres deep, which determined the length of the spuds. The decks had to be large enough to give a generous area for manoeuvring vehicles and to be able to carry 40-ton tanks.

Connecting the pierheads to the shore were four 3-metre-wide floating roadways made up of 24-metre-long interlinking pontoons called 'whales'. The longest roadway was 1.6 km long. Being subject to wave, current, and wind pressures meant the linkages between them had to be designed to withstand translatory movement along each of three axes as well as rotational movement about each of three axes. Design was based on a wave height/length ratio of 1:9, a wavelength of 50 metres, and a level difference between adjacent pontoons of 5 metres. The bridge bearings were spherical and designed to allow a free angular movement of one span relative to another of 24° and a torsional displacement

Interlinked whales. ICE

of 40° along the length of each bridge span. To allow these very large angular movements, the bearings of adjacent spans were arranged to sit one inside the other, the outer bearing only taking its support on the pontoon. Made of mild steel, each 'whale' weighed 28 tons.

To prevent the lateral displacement of the roadways, the 'whales' were moored to the seabed by very long cables each about 14 times the depth of the water below the pontoon. The tension in the cables varied from 5 tons at low water to 12 tons at high water and up to 25 tons in a storm. Investigations having shown that as traditional anchor designs were unsuitable in these circumstances a 'kite' anchor was developed. Dropped in the normal way it would burrow below the seabed with increased load and, in reasonable ground, was capable of holding loads up to 30 tons. They were laid by small plywood-constructed craft called 'mooring shuttles'. Modified arrangements had to be adopted where the seabed was rocky.

Many of the designers received awards from the Royal Commission on Awards to Inventors. In addition to those who had been involved in the design of the components, many more of Britain's leading civil engineers were involved in overseeing the assembly and other work on site.

The conveyance of troops, armaments and stores continued after D-Day virtually unimpeded by German operations, though for a period it was seriously slowed down by the gales that devastated the American Mulberry harbour on 18 June. By 5 July the one-millionth serviceman had been landed.

Beckett, Major Alan MBE RE (1914–2005)

He designed the 'whale' floating roadways that connected the Mulberry harbour pierheads to the shore and the 'kite' anchors that kept them in position. After studying civil engineering at the University of London he worked with structural engineers specialising in bridge design prior to joining the Royal Engineers in 1940. A trial length of his floating roadway was erected off the Scottish coast; after it survived a severe storm, he was ordered to arrange for the manufacture of 16 km for use off Normandy. Post-war he initially went into partnership with Brigadier Sir Bruce White, who had been in overall command of the Mulberry harbours, but later became a founding partner of the firm Beckett Rankine. In 2009 a memorial to him, together with a replica 'kite' anchor was erected in Arromanches. Two years later a stained glass window dedicated to him was installed in St Peter's Church, Oare, near Faversham, Kent, where he is buried.

Major Beckett's memorial at Arromanches with a kite anchor. Wikipedia

As a supplement to the Mulberry harbours, the planners had arranged for fuel tanks to be constructed on shore and connected by 1-km-long 150 mm diameter pipelines to buoyed flexible connections at which tankers could discharge. These came into their own in the American sector when their harbour was wrecked.

The success of the maritime operation was a tribute to the planning and resourcefulness of the British and American navies; as Correlli Barnett said in his concluding remark about Operation Neptune: 'For the Western Allies . . . seapower remained as ever the midwife of victory on land.'

Hobo's Funnies

In 1943 Major-General Percy Hobart (Hobo) was appointed to establish and command the specialist 79th Armoured Division with instructions to modify and develop armoured vehicles which could be used by the RE when storming beaches, breaching defences, and

Hobart, Major General Sir Percy KBE CB DSO MC (1885–1957)

Also known as 'Hobo', he was a British military engineer noted for his command of the 79th Armoured Division during the Second World War. He was responsible for many of the specialised armoured vehicles that took part in the invasion of Normandy and later actions. Graduating from the Royal Military Academy, Woolwich, in 1904 he was commissioned into the Royal Engineers but after service in the First World War he volunteered to be transferred to the Royal Tank Corps in 1923. Having been promoted to major general, in 1938 he was sent to Egypt to establish a mobile force which later became the 7th Armoured Division and achieved fame as the 'Desert Rats'. He was an outspoken and unconventional officer and in 1940 was dismissed and retired from the Army. Having returned to the UK he joined his local Chipping Campden Home Guard which thereupon became 'a hedgehog of bristling defiance'. Following a newspaper article about him, Churchill ordered that he be reinstated. After training an armoured division he was appointed to establish and command the specialist 79th Armoured Division. Under his leadership the division modified conventional armoured vehicles to enable them to assist the Royal Engineers in carrying out many of their operations when storming beaches and breaching defences. His sister had married General Montgomery who rapidly appreciated the contribution these armoured vehicles could make. Known as 'Hobo's Funnies', they were assigned to operate as part of other units and were described as 'the decisive factor on D-Day'. By the end of the war some 7,000 had seen service.

Duplex drive 'swimming' tank with waterproof screens lowered (upper) and erected (lower). IWM

crossing gaps. Known as AVREs (Armoured [or Assault] Vehicles Royal Engineers) and nicknamed 'Hobo's Funnies', a large number were deployed on D-Day and in subsequent actions. They made a critical difference to the success of the landings and have been described as 'the decisive factor on D-Day'. Typical of them was this 'swimming tank' which could propel itself on the surface when the waterproof skirt was raised.

Some DMWD Inventions

The DMWD conceived a number of devices which also contributed to the Allies' triumph on D-Day. Amongst them were radar markers and radar decoys, designed to guide the invasion fleet at night and mislead and confuse the enemy as to what was reflecting signals. They also developed 'Swiss Roll' floating roadways (rolled up for transport to the site and used by the Royal Navy for landing its stores and personnel on the beach), and rocket landing craft, which allowed invasion forces to fire at enemy positions during the interval between the end of the naval bombardment and the actual landing. These proved highly successful in both the Mediterranean and at Normandy and were regarded by some as the most valuable of all DMWD's projects.

Other DMWD inventions included hedgerow landmine detonators which were developed from the Hedgehog depth-charge launcher, the essential links being the forward-launch concept and the fuse needed to detonate the explosives at the right instant. Mounted on landing craft, they fired their charges as the craft approached the beaches so as to detonate any mines on the beach. They had not been a success in the Mediterranean, but a number proved their worth in Normandy. For scaling the 30 metre coastal cliffs at Pointe du Hoc during the Omaha beach advances rocket-propelled grapnels were used, as were extendable ladders which had been adapted from London Fire Brigade fire escape ladders and mounted on landing craft.

Inventions used in connection with the Mulberry harbours included the helter-skelter chute to speed the transfer of men from troopships to landing craft. Floating Bombardon breakwaters were also tested and in part developed by the DMWD, whose engineers realised that waves were confined to the surface and thus could be dampened down by a floating obstacle. They also developed the alternative Lilo breakwater which was based on the principle of an inflated balloon.

Communications

As with the activities of the other services and corps, the preparations for the RSigs' participation were part of an inter-Allied tri-service operation. The joint naval, army, and air force planning unit had to consider how best to establish the communication links and other support services that would be needed by the infantry and armour, as well as by the Royal Navy and RAF during and following D-Day.

As well as more conventional communications equipment, submarine cables were ordered; methods of transporting them to, and speedily connecting them on, the Normandy beaches were developed, as were techniques for rapidly establishing radio links. While many lessons from Italy were applied, the improvised nature of some of the systems adopted there rendered them not wholly appropriate for Normandy.

Despite some rough seas, very few signal vehicles and equipment were lost during the crossing. Moreover, thanks to the efficiency of the waterproofing, the equipment mostly functioned satisfactorily on arrival. Small, light, truck-mounted transmitters arrived with the early formations and were used onshore until it was possible to land and operate heavier trucks with stronger transmitters. Despite the large number of channels and systems that eventually became operational there was relatively little mutual interference.

The main centres between which communications (teletext, telephone, and radio) were provided included Supreme Allied Headquarters in Westminster, the large administrative centre at Southampton, the American centre at Plymouth, the British centre at Portsmouth, the communications headquarters at Dover, and terminals in the bridgeheads in France.

It was a measure of the success of the planning and associated training of the troops involved that effective communications were established soon after the landings and that the events that actually happened were anticipated reasonably accurately with few of the reserve resources needing to be used.

Among the highlights were the airborne forces' communication systems and the radio links between the assault troops on shore and the offshore command vessels

Mine clearing flail tank. By kind permission of The Tank Museum

Bobbin Tank. IWM

Fascine-carrying tank. IWM

Ark Tank. IWM

which were established within a few hours of alighting. The first cross-Channel radio link from Normandy to the Isle of Wight was operational one day after D-Day, and three days later there were 24 cross-Channel radio links operational.

The laying of the first submarine cable began two days after D-Day from Southampton; it was operational two days later, and by the end of August four submarine cables were providing 27 speech and 39 telegraph channels. In France the links were based in appropriate communication centres.

Users of the radio transmissions were warned that most of their messages were liable to enemy interception and thus had little security value, and that messages should be construed accordingly.

REME Operations

The Corps had been investigating the provision of lightweight vehicles that could be carried in air- and landing craft, and they had also developed means of waterproofing the equipment. During the fighting REME personnel worked valiantly to move ordnance, vehicles, and other equipment causing obstructions; service tanks and transport vehicles; replace or repair engines, tracks etc. on fighting and other vehicles; and arrange for the return to the UK of engines and other equipment needing major maintenance. With 6,000 armoured vehicles and 200,000 transport vehicles landed, the size of the task can be imagined. For this, REME Beach Recovery Sections were formed, each being responsible for 1 km of beach, and a Beach Armoured Recovery Vehicle (BARV) was developed which could operate in water more than 2 metres deep. Frequently, 'controlled cannibalisation' of equipment was the preferred solution. REME personnel also had the gruesome task of removing the corpses remaining in damaged tanks that were to be restored.

The amphibious American-built DUKWs that were used to transfer stores from ships lying offshore to onshore dumps worked well at sea but were frequently damaged by the inadequate roads and tracks that they had to use onshore, and this caused major recovery problems.

Enemy mortar fire caused considerable damage to Allied troops and vehicles during the fighting in the Bocage country of Normandy; despite the Army's best efforts, the mortars' locations were discovered only when a young REME officer attached to an anti-aircraft regiment noticed that the tracks of the missiles could be seen on his semi-mobile radar set.

The Advance into Europe

After fierce fighting for a number of weeks the Allies broke out from the bridgehead and started their advance through north-east France, liberating Paris on 19 August. Following more tough battles over the subsequent months they were eventually able to liberate the Low Countries and enter Germany.

BARV recovering stranded lorry. http://worldwar2headquarters.com/HTML/normandy/HobartsFunnies/hf_BARV.html

Unloading an LST at a beachhead. From the Collection of the USS LST Ship Memorial.

DUKW proceeding shoreward

The RSigs and REME, together with their American colleagues, respectively established sophisticated communications networks and recovered and rapidly repaired any armour or ordnance that had been damaged in the fighting.

As in Italy, one of the principal tasks for the RE and the US Corps of Engineers was to provide bridges for crossing rivers and other obstacles. In north-west Europe the terrain was less mountainous than in Italy and, particularly in the Low Countries and Germany, pontoon bridges were often chosen because of the wider rivers and canals that had to be crossed. Many were over 1 km long and had to be constructed under conditions of flood and snow as well as enemy fire. Their other activities included finding and defusing mines and booby traps, and constructing dozens of temporary airstrips so that the Allied air forces could provide close support to the armies

A Class 40 Bailey bridge was built across the River Orne at Caen shortly after securing the bridgehead. Prior to the assault on Le Havre, the RE built two Class 40 Bailey bridges and a Class 9 folding boat bridge over the River Seine at Vernon, while a scissors bridge was laid over the River Somme and some existing bridges were captured intact. This enabled a rapid advance of some 400 km in six days across northern France and into Belgium, progress further aided by the capture of more undamaged bridges.

Thought had been given as early as 1942 to the crossing of the Rhine, and a trial construction was rehearsed in the UK to ensure that the right equipment would be available. First, however, the Maas and Maas-Waal Canal had to be crossed, and 11 bridges, including two Class 70 and three Class 40 bridges, were built over the river, as well as a number over the canal.

After the first troops and armour had been ferried across the Rhine by Buffalo AVREs and pontoon rafts, bridging work began on eight pontoon bridges and two that used folding boat equipment. The 330-metre-long pontoon bridge at Xanten was the first to be completed; it was operational within 31 hours of the work beginning, and it ensured that nearly 30,000 vehicles could cross within six days. All the bridges were completed within ten days of the first assault.

A Bridge Too Far

Immortalised in the film (and phrase) *A Bridge Too Far*, Operation Market Garden was an airborne assault aimed at capturing four adjacent bridges over major canals and rivers near the Dutch-German border between Arnhem and Wesel. The operation failed because many troops were dropped too far from their objectives, the main relief force arrived too late to be effective (largely because of inadequate roads), and two German Panzer divisions happened to be retraining in the area at the time. The RE was heavily involved, with sappers of the 6th Airborne Division being eventually forced to fight as infantry in hand-to-hand engagements.

Bridge over the Maas-Waal canal (by permission of the Institution of Royal Engineers).

330 metre Bailey pontoon bridge over the River Rhine at Xanten (by permission of the Institution of Royal Engineers).

After crossing the Dortmund-Ems Canal, three bridges were built over the Weser, one as a replacement for a bridge that had been abandoned prior to completion due to significant casualties and equipment damage. An advance of some 160 km in two weeks was made after a dozen bridges were repaired and some 30 major bridges were built, including one over the 280-metre-wide River Elbe at Artlenburg, which was completed three days prior to the German surrender.

Semi-permanent Road and Rail Bridges

Like in Italy, as the front advanced it became possible to construct more permanent and substantial bridges in the rear areas, an activity which continued for many months following the cessation of hostilities. Among these were a number that were built across the Rhine and other major rivers; Major Ralph Freeman RE was entrusted with their conception and overall design. They were mostly supported by Bailey piers built from American pontoons and comprised a number of spans, some of which could be disconnected and floated into nearby docks when the bridges were threatened by floating ice.

One such bridge was the 600-metre-long Freeman Bridge (a name chosen as a tribute to Freeman, much to his surprise), whose design considerations included allowance for severe icing conditions, a water level variation of 9 metres, a maximum current of 13 km/h, and the provision of a navigational opening providing 8.8 metres clear height and a width of at least 30 metres. A further two were the twin Class 70 and Class 40 1,500-metre-long Tyne and Tees bridges across the Rhine at Xanten and Rees respectively (and were the longest military-built bridges anywhere in the world), and the pier-supported semi-permanent Class 40 1,200-metre-long Dempsey.

600-metre-long Freeman Bridge. ICE

Freeman, Sir Ralph CVO CBE FICE (1911–98)

He was the eldest son of Sir Ralph Freeman (Senior), a noted bridge engineer who had designed Sydney Harbour Bridge and was the senior partner of engineering consultants Freeman Fox and Partners. After graduating from Oxford Ralph (Junior) joined steel fabricators Dorman Long & Co. and worked on bridgework in the UK and southern Africa. In 1939 he joined Freeman Fox and Partners and worked on various military construction projects prior to entering the Royal Engineers in 1943. At the Experimental Bridging Establishment he adapted Bailey bridge components for use on a 120 metre suspension bridge which was later used in Burma. Following the Normandy landings he advised on bridge designs for crossing rivers in north-west Europe and played a key role in the design and erection of three high-level Bailey bridges across the River Rhine, one of which was named the Freeman Bridge. He was co-author of the paper on 'The Erection of Military Road Bridges, 1939–46', presented at the 1948 conference on 'The Civil Engineer in War'. Post-war he served in the Territorial Army Volunteer Reserve, and commanded the prestigious Engineer and Railway Staff Corps 1969–1974.

Following demobilisation he returned to Freeman Fox and Partners; appointed a partner in 1947 and senior partner in 1963, he was responsible for many major post-war projects. These included: the Dome of Discovery for the 1951 Festival of Britain; the M5 motorway; power stations in South Wales; a hydroelectric project in North Wales; radio telescopes in Canada and Australia; the Hong Kong cross harbour tunnel; the Hong Kong metro system; and bridges across Auckland Harbour, the River Clyde, the Firth of Forth, the Bosporus, and the Severn and Humber estuaries. The latter, with a span of 1,410 metres, was for 17 years the longest single-span suspension bridge in the world. He was appointed consultant to the Sandringham Estate, referring to himself as 'the Queen's plumber'.

He was president of the Institution of Civil Engineers 1966–67 and also served as chairman of the Association of Consulting Engineers, and a member of the Royal Fine Arts Commission. He was knighted in 1970 for services to engineering. His son Anthony would have become the third Freeman to be a noted bridge engineer but tragically died following an accident while inspecting a bridge across the River Tagus in Lisbon in 1997.

Navigation bay of Freeman Bridge after launch on to floating piers. ICE

Most of these bridges remained operational for many years and played a vital role in the post-war West German economic recovery. Many assault bridges were replaced and rail bridges reinstated so as to restore the railway network

The Air Campaign

The Allies' strategic air offensive in Europe is described in Appendix 5, its leaders (erroneously) believing that their efforts would on their own induce the enemy to surrender. As this had not happened by 1944 their main objective both during and after the invasion was to dominate the skies over north-west Europe in support of the Allied armies. This meant discontinuing the strategic air offensive against German military capabilities for a few weeks.

Just as the air forces supported the land operations, so the successful land operations benefited the air campaign. As the Allies advanced eastward following the break-out from the bridgehead, temporary landing grounds were constructed. The launch sites for V1 bombs and V2 rockets were forced further away from the UK while land-based stations providing signals for bombing aids could be sited nearer Germany, thereby increasing their accuracy.

Also, as suitable terrain was liberated, temporary runways were constructed using PSP (pierced steel plank), developed by US engineers, and PBS (prefabricated bituminous surfacing), developed primarily by the Royal Canadian Engineers in conjunction with the Air Ministry.

The main Allied air units delivering close air assistance to the land operations were part of the Tactical Air Force. In the main both Bomber Command and the US 8th Air Force continued with their strategic air offensive, although their resources were diverted to aid the land forces during the build-up to and the invasion itself, throughout the attacks on the V1 and V2 launch sites, and on other odd occasions.

The Allied domination of the skies was the result of a number of factors, including the Ultra decrypts provided by Bletchley Park. During the 'Big Week' of 20–25 February the USAAF targeted and wreaked havoc on the centres of the German aircraft industry. German fighters came off worst in battles the *Luftwaffe* fought against the Packard-Merlin Mustangs that were by then protecting the US bombers attacking Germany. Between January and April the Germans lost over a thousand pilots, including some of their best commanders. Their High Command admitted that 'the time has come when our weapon is in sight of collapse'.

In overall command of the Allied air forces was Air Marshal Sir Arthur Tedder, the Deputy Supreme Commander. Under him in command of operations in France on the British side were Air Marshals Leigh-Mallory, Coningham, and Broadhurst; alongside him in charge of Bomber Command was Air Marshal Sir Arthur Harris. Major Generals James Doolittle and Carl Spaatz were in command of the US 8th (fighter) and US 9th (bomber) Army Air Forces respectively and thus had overall responsibility for American air operations in Europe. Close support of ground operations was provided by the Tactical Air Force which mostly comprised fighters and fighter-bombers. Aerial photographic reconnaissance played a vital part in gaining knowledge of the situation behind enemy lines and was undertaken by unarmed Spitfires and Mosquitoes.

During the days and weeks leading up to the invasion an increased air offensive had been mounted onto targets in northern France. However, more attacks were aimed on the Pas-de-Calais and other areas away from the actual invasion beaches than on the proposed invasion areas.

In the early months of 1944 the Allied bomber commands were continuing to execute their strategic bombing campaign on targets in Germany, their commanders believing that these campaigns would shortly 'finish the war on their own'. Yet they were directed that from 31 March their targets would be those that would best support the forthcoming invasion. In determining what these should be, Tedder ordered that they should follow the 'Zuckerman Plan', the brainchild of Professor Solly Zuckerman, the scientist who had previously worked in Combined Operations and with Tedder in the Middle East. See bionote in Chapter 1. His plan was that the principal targets should be the continental transportation networks, in particular French and Belgian road junctions and locomotive depots and repair yards. A consequence of this was that there would inevitably be collateral damage and casualties among the French and other civilian populations. This gave serious concern to Churchill and the War Cabinet, but after consultations with Roosevelt it was agreed that the policy should be confirmed.

Such a strategy would seriously impede the German ability to move reinforcements and supplies, although it also had the benefit of not disclosing the proposed site for the invasion. In particular it was directed that the rail network within a 200 km radius of Normandy should be immobilised for at least two weeks after D-Day, but that the part of Normandy where the landings were to take place should not be included until D-Day-1. Thereafter all communications systems in the area were to be targeted.

A further aim of the air campaign was to deceive the Germans as to the chosen site. For every reconnaissance mission flown over Normandy, two were flown over the Pas-de-Calais; for every ton of bombs dropped on coastal batteries west of Le Havre, two were dropped north of the port; and 95% of the railway targets were north and east of the River Seine.

British and American heavy bombers participated in the attacks but were withdrawn after D-Day. All local air support thereafter was tactical and undertaken by fighter-bombers and fighters. An unexpected bonus of the switch in Bomber Command's targets was that because of the need to be more precise as to where bombs were dropped, new and improved technical aids and more powerful bombs were developed.

The campaign proved very successful and resulted in a creeping paralysis of the rail network, with those trains that still ran moving very slowly and travelling only at night – and often having to make long detours. The bombing of the railway network continued after D-Day. On the night of 8–9 June for instance, the main line from south-west France to Normandy was targeted where it crossed a bridge and then entered a tunnel at Saumur, the 5,000 kg Tallboy bombs entering the ground so fast that the noise of their fall was heard after the explosions.

General Montgomery's plan for the invasion included the capture of Caen at an early stage, though for a variety of reasons this was not achieved until six weeks after D-Day. Thus the Allied air forces could not use its airfield or the flat ground southeast of the town during the critical early phases of the invasion. Instead they had to rely on airfields in southern England and a number of temporary airfields – in Sussex, for example, eight temporary advanced landing grounds were built to supplement the six permanent airfields in the county.

The airborne invasion forces totalled some 17,000 parachutists and glider-borne soldiers who were carried or towed by 71 squadrons of aircraft. The first landing on French soil at 0020 hours on D-Day was by six Horsa gliders carrying men of the 2nd Battalion of the Oxford and Buckinghamshire Light Infantry to secure what became known as the Horsa and Pegasus bridges over the River Orne and Caen Canal.

In many earlier campaigns, the Allies had suffered disastrous losses when they did not have air supremacy. By 1944 the lesson had been learnt, and many of the new weapons and innovations introduced had gone into ensuring that in Normandy they would have total command of the air. This was undoubtedly one of the principal factors that made the Normandy invasion so successful.

Harris, Sir Alan CBE (1916–2000)

The son of an Admiralty electrical engineer who was born in the West Country but brought up in London where, while working for Hendon Borough Council, he studied for five nights a week at what is now City University to obtain a degree in civil engineering.

During the Second World War he served with the Royal Engineers, later saying of his time in the forces that he would not have missed a moment of it. He qualified as a diver and served as an officer in a port construction and repair company. He landed in Normandy on D-Day +1 and, working from a small fleet of French fishing boats, commanded diving operations, including mine clearance, at Mulberry B harbour at Arromanches. Later he moved to Ostend where he first met the engineer–inventor Gustave Magnel, who told him about prestressing concrete. He then moved to the Rhine where he worked on constructing bridges before demobilization in 1946 with the rank of major. He was awarded the Croix de Guerre and mentioned in dispatches. Post-war he served with RE Territorial Army units where he attained the rank of colonel.

After the war Harris persuaded Eugène Freyssinet (the principal pioneer of prestressed concrete) to give him a job in France and, as a result, Harris worked on many reconstruction schemes that involved prestressed concrete including bridges, harbour works, a runway at Orly as well as more mundane projects, such as precast concrete flooring units and railway sleepers. His relationship with Freyssinet was similar to that between an apprentice and his master. An enthusiastic Francophile, he gained a good working knowledge of French – and had an amusing habit of 'Frenchifying' an English word if he did not know the appropriate French word.

He returned to the UK in 1949 as head of Freyssinet's British subsidiary but in 1956 he established his own consultancy in partnership with his brother John and James Sutherland. In addition to exploiting their expertise in prestressed concrete the firm designed many significant projects and had offices across the UK and in Australia, Hong Kong and Singapore. Harris was elected President of the Institution of Structural Engineers and vice-President of the Institution of Civil Engineers and appointed a professor at Imperial College. He was a great raconteur and most amusing after-dinner speaker. He was knighted in 1980 and, in addition to his British honours, he was also awarded the Ordre du Mérite by the French government[1] in addition to the Croix de Guerre.

1 Institution of Structural Engineers; DNB; New Civil Engineer

RN Clearance party at work. http://www.mcdoa.org.uk/rn_clearance_diving_branch.htm

Mines and Booby Traps

As well as the mines laid in harbours, retreating German forces destroyed bridges and cratered road junctions and airfield runways etc. They also mined and/or booby-trapped bridge approaches, harbour installations, communications centres, and other facilities that the Allied armies would want to use.

Onshore, the immobilisation of these devices was the responsibility of the RE and RAF engineers, while specialist inter-service mine clearance teams worked in harbours and rivers. One such unit under the command of Major (later Colonel Sir) Alan Harris RE comprised seven sappers, a Royal Marine, and three naval divers.

One of the novel gadgets developed by the DMWD and used in these operations was the wreck dispersal pistol, which could simultaneously detonate a number of explosive charges attached to underwater obstructions (and for which Lt Cdr A. C. Brimstead RNVR was awarded £1,500 by the Royal Commission on Awards to Inventors). Another was the 'Kapok Jacket' developed in conjunction with the RN Medical Branch to protect frogmen working to defuse unexploded mines against accidental explosions.

After D-Day: Naval Operations

During the advance eastward by the Allied armies the RN and United States Navy continued with the vitally important tasks of ensuring the safe passage of troops, equipment, and stores from the UK and USA; minesweeping the approaches to the Normandy beaches;

and protecting the flanks of the maritime operational area. Furthermore, as ports were liberated, they had to clear mines from the harbours and estuaries and, in conjunction with the Army, reinstate cranes and other dockside facilities.

After the invasion, U-boats based in the French Atlantic ports attempted to attack Allied vessels; however, 12 U-boats were sunk in the Channel and Bay of Biscay and 13 more elsewhere in June alone. Similarly, the few German destroyers and E-boats that ventured forth were rapidly disposed of.

Within the protected maritime area, transport vessels quickly established a smooth cyclical rhythm of discharging their cargoes, returning to the UK to replenish, and then sailing once more to Normandy to repeat the cycle; by the end of August some two million troops had been conveyed to France.

As with all other ports that had been occupied by the Germans, the dockside facilities at Cherbourg were destroyed, and it took some weeks to make the port serviceable. In early October operations at both Cherbourg and the Mulberry harbours were temporarily interrupted by a gale, but the situation was partly relieved in mid-October by the capture of Marseilles and Toulon, whose port facilities were soon put into service despite being in the south of France.

The Army's advance eastward, coupled with the growing size of the Allied armies, highlighted the urgent need to secure a working port relatively near the front line through which supplies could be landed. In the autumn this became a top priority for Supreme Headquarters. After extremely heavy fighting, Le Havre was liberated on 12 September and Brest on 19 September, but the Germans had deposited oyster mines and sunk blockships in both harbours, thus making them unusable until they had been cleared. As both ports were away from the front, neither was used until after the cessation of hostilities. Rotterdam was not freed until April 1945 and it too was not operational while the fighting continued.

Antwerp was liberated on 4 September and it was decided that it should become the main Allied port. A directive of 16 October from General Eisenhower to General Montgomery stated: 'Operations destined to open the port (of Antwerp) will therefore be given complete priority over all other operations.'

It was, however, 130 km up the Scheldt Estuary which had been heavily mined by the Germans, who also still occupied the islands of South Beveland and Walcheren at the head of the estuary. This resulted in a mini-Normandy amphibious assault by the British and Canadian armies on the two islands; after heavy fighting, they were successfully taken in early November. The clearance of mines from the estuary was then essential, and ten squadrons of minesweepers were deployed. Eventually, on 28 November, some two months after Antwerp had been liberated, a convoy of 19 Liberty ships went alongside the quays, after which Antwerp became the main supply port for the Allied armies.

New Type of U-boats

Although the Allied navies had essentially defeated the menace of U-boats, the Germans were, with great difficulty, still building large numbers of submarines. With *Schnorchel* tubes and radar-detecting sensors, they were proving more difficult to locate and attack, especially as they were now operating singly rather than in wolf packs. Also, they were mainly operating around the British coast where they could hide amongst the wrecks that littered the seas and where the temperature layers created by tidal streams and river estuaries often made sonar detection difficult.

As the First Sea Lord reported in September,

> *The submarines and the oyster mines have been giving us considerable trouble … Our escort groups have had to learn a completely new technique … [as] they are so dispersed that it is actually taking a greater effort than when they went about in packs in the Atlantic.*

The Admiralty's response was to increase the number of ships and aircraft hunting the U-boats, a tactic which largely proved successful.

New weapons were also introduced, including the Squid Mortar, the aerial acoustic torpedo, and the retro-bomb which was fired behind an aircraft that had just overflown a submarine located by a magnetic anomaly detector. Rocket-firing Typhoons joined Coastal Command's Mosquitoes and Beaufighters in attacking U-boats, 27 of which were sunk in the last five weeks of the war alone.

Equally worrying for the Admiralty were the new types of submarines (Type XXI and the smaller Type XXIII) that the Germans were developing and producing. These had streamlined hulls, greatly enlarged batteries, and very powerful electric motors. They therefore had submerged speeds comparable to those of surface vessels and could cruise without refuelling for longer distances than conventional U-boats. The hull sections were built in separate yards and transported by barge along rivers and canals to a coastal assembly yard. But the Allied aerial bombing campaign was now causing such havoc to Germany's transportation system and industrial production that the first Type XXI became operational only in the month before the end of the war.

By the time of the German surrender only two of her capital ships were undamaged; 221 submarines scuttled themselves and the remainder, including some Type XXIs and Type XXIIIs, sailed under escort to British ports.

Sadly, Admiral Ramsay, the mastermind of the naval side of the Normandy invasion, as well as of the Dunkirk evacuation and the Sicily operation, was killed in an air crash in January 1945 and so did not live to see victory achieved.

Hartley, Arthur CBE (1889–1960)

The chief engineer of Anglo-Iranian Oil Company (now BP), and involved in the development of the FIDO fog dispersion system installed at some Bomber Command airfields, the PLUTO and DUMBO submarine fuel pipelines and the bombsight which sank the Tirpitz. A graduate of Imperial College, London he worked with the North East railway company and an asphalt company before joining the RFC in WW1. He qualified as a pilot and joined the Air Board where he was involved with the development of interrupter gear. After the war he spent five years as a consulting engineer before joining Anglo-Iranian. After WW2 he returned to Anglo-Iranian until his retirement in 1951 when he was elected President of the Institution of Mechanical Engineers. He was elected President of the Institution of Civil Engineers in 1959, but died three months into his tenure.[1]

1 Institution of Civil Engineers, *Obituary*, April 1960

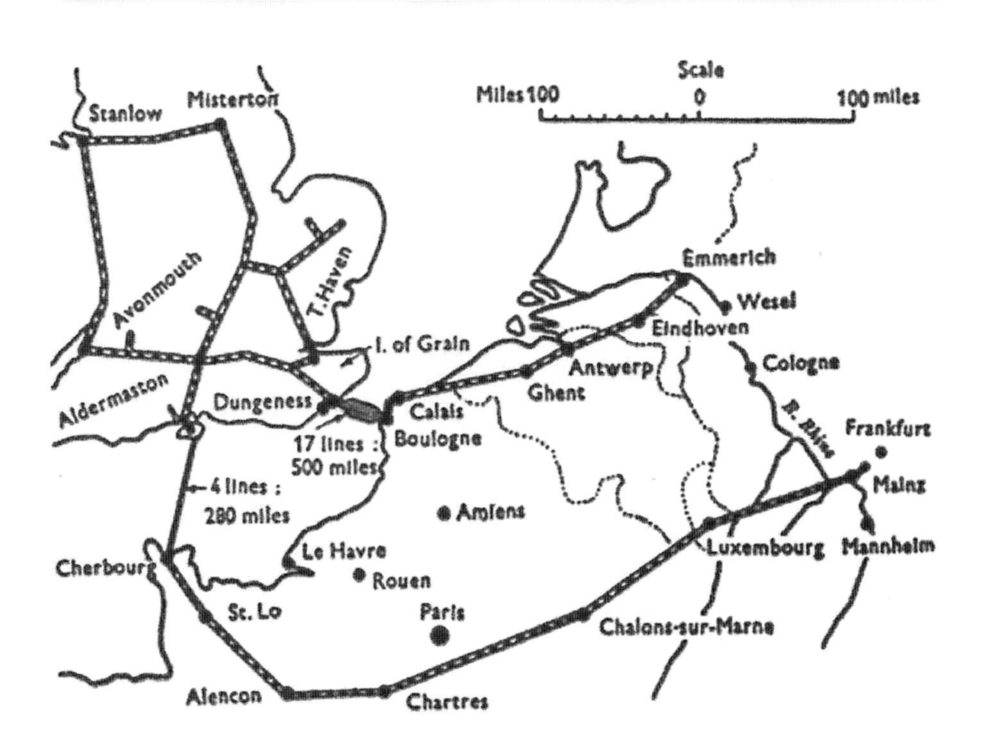

The complete network of pipes. ICE

PLUTO and DUMBO

The requirement for cross-Channel submarine pipelines to supply fuel to the Allied armies in Normandy was appreciated by the military in April 1942. As a result Arthur Hartley, the chief engineer of Anglo-Iranian Oil Company (now BP), was approached the same month. While the concept was not new, a complication was that vessels joining adjacent lengths of pipeline would be stationary and therefore easy targets for the enemy. Thus there was a need for an alternative solution which did not require the joining of pipelines at sea and this was attained by the development of extremely long pipelines.

Under the acronym PLUTO (Pipe Line Under The Ocean) the original intention had been for a short pipeline across the Strait of Dover; however, it was later decided that longer pipelines running from the Isle of Wight to Cherbourg would be preferable. Because of prolonged German resistance in that port, the Allied armies were advancing eastward by the time it had been liberated, and it was then agreed that further pipelines should be laid from Dungeness in Kent to Boulogne in France. These were named DUMBO (Dungeness to Boulogne), and as the Allies moved further through the Low Countries and into Germany, they were extended as far as Germany itself. In the UK they connected with a 1,600 km internal pipeline network running from Merseyside and Avonmouth to South East England. To avoid detection by aerial reconnaissance the network was constructed at night, with pumping stations in buildings with innocuous external appearances.

Two heavily reinforced types of pipeline were used – a seamless lead pipe named HAIS 'cable', so named for security reasons, and the HAMEL steel pipeline. Extensive pressure testing and trial laying of both pipelines were carried out before the final designs were determined. HAIS was originally conceived as a 5 cm diameter lead pipe but a 7.5 cm internal diameter pipe with an external diameter of 8.7 cm was eventually adopted, and a total of 1,300 km was manufactured, some of it in the USA. It was made up of lead pipe, two layers of compound paper tape, one layer of bitumen-prepared cotton tape, four layers of steel tape, one layer of jute bedding, 57 galvanised mild steel armouring wires, and two layers of jute. The HAMEL pipeline was of 7.5 cm internal diameter and had the advantage that adjacent lengths could be welded.

For laying the pipelines on the sea bed, and to avoid having to join lengths of pipe at sea, a large bobbin-like 9-metre-diameter drum was constructed on a barge (nicknamed HMS *Conundrum* – cone-ended-drum). After successful trials a number of other vessels were similarly converted. Over 100 km of pipeline could be loaded onto a drum. A special naval task force, Force Pluto, was established to undertake the marine operations. Three merchant vessels were requisitioned and modified so as to be able to tow the drums. Eventually four 130 km pipelines (2 HAIS and 2 HAMEL) were laid from the Isle of Wight to Cherbourg, and 17 (11 HAIS and 6 HAMEL) from Dungeness

Laying the pipeline: A 'Conundrum' being moved into position into a specially constructed dock in preparation for the winding on of the pipe. IWM

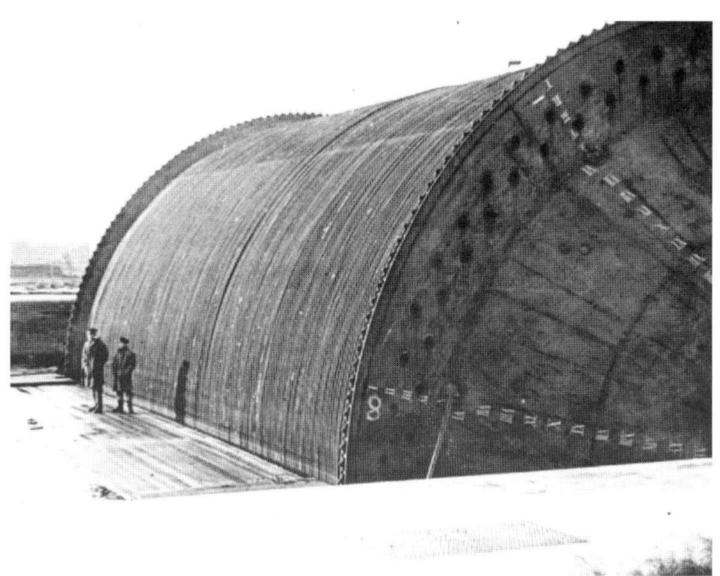

A 'Conundrum', loaded with pipe, ready to be towed across the Channel. IWM

A 'Conundrum' being towed across the Channel laying out pipe behind it. IWM

to the Pas-de-Calais. These were the forerunners of the flexible pipelines used in today's offshore oil fields.

Unfortunately, there were technical problems with the four pipelines to Cherbourg and no products were passed through them. On the other hand, the first DUMBO pipeline became operational in early December when 600 tons per day of petroleum were pumped through it, and by March 1945 3,000 tons per day were being pumped through the 11 pipelines

After D-Day: Communications

After the break-out from Normandy, some of the links to the UK were extended to work by relay direct to Army headquarters, and a priority 'red line' radio service was instituted between Supreme Allied headquarters and the Army commanders. Initially, because units were operating closely together, problems arose because of interference from the signal traffic of adjacent units, thus requiring careful allocation of the frequencies used. (Later, the lines of communications reached up to 500 km, which raised its own set of challenges.) Also, due to the large volume of traffic, special measures had to be adopted to give priority to vital messages. As territory was liberated, the restoration of the civilian networks was a main priority, although this was complicated by the Germans having soaked some of the repeater stations in petrol and burning them, hence making it necessary for mobile stations to be built. Eventually, two-thirds of the communication services were provided by a restored civilian network and one-third by new landlines.

On occasions the Army's rate of advance was extremely rapid and the landline operations could not keep up with it. In these circumstances mobile radio link equipment was invaluable in providing a few essential circuits, some of which used as many as six relay stations resulting in a link having seven legs each up to 40 km long.

Other measures which were introduced included a cross-Channel dispatch boat service operated by the RN, with sorting offices at each terminal and eight special signals regiments which were used for communications with the RAF. The supply of stores from signal parks was controlled by an ordnance officer.

For taking communications cables across major rivers it was found not to be practicable to attach the cables to pontoon bridges, partly because of their instability but also because they would interfere with the repair or replacement of damaged pontoons. Instead cables were suspended from the piers of demolished bridges, laid on the riverbed, or, for short crossings, suspended from shore to shore.

In three major airborne landings special measures had to be adopted for internal communications and links with appropriate headquarters. These were satisfactory in support at Normandy and the Rhine, though they were less successful at Arnhem.

The Royal Signals' official history concluded its review of the operations in north-west Europe by saying:

The range, intensity and tempo of the operations were indeed formidable, and the problems which confronted Signals gave them but little respite or cause for complacency. It is therefore a matter of considerable satisfaction that the organisation, equipment and methods employed were able to surmount these exacting demands.

After D-Day: REME Operations

When the preparations for assaults across rivers were underway, two of the Corps' major tasks were waterproofing vehicles that were about to cross rivers other than by bridge and undertaking major servicing of armoured vehicles in specially dedicated workshops so as to ensure that they would be battle-worthy. Throughout the Rhine and other crossings they also had to: provide an adequate recovery organisation for keeping entrances and exits clear at each crossing point; maintain organisation in the marshalling areas in order to keep the roads and bridges clear; carry out minor repairs to vehicles; and after a crossing, arrange the provision of workshops as close as possible to the advancing troops.

Some REME personnel were trained by the Royal Navy in shallow-water diving so that they could help with the removal of both disabled light naval craft and 'drowned' vehicles that had been damaged before reaching the beaches.

The nine Anti-Aircraft Brigades in the British Expeditionary Force each had REME units attached, and because enemy air activity was relatively limited, AA guns were sometimes used as field artillery or in an anti-tank role. Also, when V-1 attacks on Brussels and Antwerp began, special workshop units were sent from the UK in support of the Anti-Aircraft Artillery Regiments deployed. Army reconnaissance and other aircraft were supported and serviced by the RAF.

Representing some 4.7% of the total force in Europe, REME strength peaked at about 36,000, of whom nearly 7,500 became casualties, and of these 1,500 were battle casualties.

VE Day

The Western Allies' advance into the German mainland coincided with the Russian advance from the east and resulted in the German unconditional surrender, which became effective on 8 May.

The five years since the British evacuation from Dunkirk and the near four years since the German invasion of Russia had seen a remarkable transformation of capabilities and fortune – in no small part made possible by the technological advances developed by the scientists and engineers of the Allied nations.

But, on the other side of the world, the Japanese were still fighting as tenaciously as ever, and continued to do so until technology brought the war to a sudden end...

SOUTHEAST ASIA AND PACIFIC TIMELINE

1931

Japan invades Manchuria (a region of north-east China).

1937

Japan occupies regions of eastern China. Chinese resist Japanese expansion.

1940

Japan invades French Indo-China (now Vietnam, Laos, and Cambodia).

1941

7 Dec Japanese naval aircraft attack US Fleet at its base at Pearl Harbor, Honolulu.

8 Dec Japan attacks Malaya, Hong Kong, Thailand, and Philippines.

10 Dec HMSs *Prince of Wales* and *Repulse* sunk by Japanese aircraft.

25 Dec Hong Kong surrenders.

1942

Jan Japanese invade Burma (now Myanmar) and Dutch East Indies (now Indonesia).

15 Feb Singapore surrenders.

19 Feb Japanese aircraft bomb Darwin, Australia.

27 Feb Battle of Java Sea (Japan beats combined US/UK/Dutch/Australian fleet).

7 Mar British evacuate Rangoon, Burma.

18 Apr US naval planes attack Tokyo.

US General Stilwell appointed the Allies' commander in chief China – marches from Burma to India, and moves into China.

4 May Japanese and US fleets fight inconclusive Battle of Coral Sea (off Australia).

31 May Abortive attempt by three Japanese midget submarines to sink ships in Sydney Harbour.

Jun Japanese Fleet submarines attack shipping off NSW coast and bombard Sydney and Newcastle.

4 Jun US fleet defeats Japanese in Battle of Midway Island.

Aug US captures Guadalcanal, Solomon Isles, and starts Pacific campaign to recover Japanese occupied islands.

SOUTHEAST ASIA AND PACIFIC TIMELINE

1942 (cont.)

	USAAF establishes 'The Hump' to fly US arms over the Himalayas into China (to replace the inaccesible Burma Road).
Dec	Abortive British attempt to recover Akyab Island on Burmese Arakan coast.

1943

Feb	First Chindit campaign; 3,000 Special Forces operate behind Japanese lines in Burma.
2 Mar	Battle of Bismarck Sea – US and Australian aircraft annihilate convoy heading for New Guinea.
4 Sep	Allies start offensive to retake New Guinea.
Aug	British Southeast Asia command reorganised and reinforced.

1944

Jan	US forces assault Marshall Islands as part of the campaign to recover control of the Pacific.
Mar	Japanese forces in Burma invade north-east India, but repulsed at Kohima and Imphal.
	Second Chindit campaign; 20,000 strong British force starts operations.
	British and Chinese forces start to enter west and north Burma respectively.

1945

Jan	US land on Luzon, Philippines.
	Ledo Road from India to China reopened.
May	British airborne forces land south of Rangoon and capture city.
Jul	Philippines liberated.
6 Aug	Atomic bomb lands on Hiroshima.
9 Aug	Atomic bomb lands on Nagasaki.
14 Aug	Japanese surrender.
15 Aug	VJ Day.
28 Aug	Japanese Army in Burma surrenders.
	Major British naval and land forces land in Rangoon and start recovering Burma.
30 Aug	British recover Hong Kong.
5 Sep	British start recovering Malaya.

Chapter 10

SOUTHEAST ASIA AND THE BELLICOSE PACIFIC

The final surrender

Few theatres of war proved as challenging to Allied engineers as the Far East, not only because of the distances involved but also because of the hostile environmental conditions in the tropical jungles of Burma (now Myanmar). The British Army formations were now called the XIV Army and commanded by General William Slim.

Since most of the few bridges that existed in Burma were destroyed when first the British and later the Japanese were retreating, the reconstruction of the roads and bridges became a top RE priority during the campaign to recover the country. Also, the relatively little basic infrastructure in the country meant that the temporary works needed to enable a project to be constructed were frequently as great a challenge as the project itself.

A prime example of this was the 333-metre-long Grub pontoon bridge across the Chindwin River, located nearly 500 km from the nearest railhead. This meant that the logistical effort and temporary works required to transport the equipment to the site were prodigious, the latter involving 22 timber and nine Bailey bridges over the final 16 km stretch alone. Once everything was in place, the bridge itself was built in five days in December 1944 by Indian Army engineers; it was the longest floating Bailey bridge constructed in any theatre of war until the final days of the European campaign. Where roads ran on steep cliff faces it was frequently necessary to use Bailey bridges to construct catwalks on the escarpments.

Another noteworthy bridge was the 135-metre-long Bailey suspension bridge over the River Shweli near the Sino-Burmese border. This was erected by the US Army Corps of Engineers in March 1945 and was the longest suspension bridge built during the war.

After the Army had stabilised its position in north-east India, now partly Bangladesh, the RE began the construction of a variety of projects, many of which were in the most inhospitable terrain. Amongst them were roads and other basic infrastructure in north-east India; some 800 km of oil pipeline; the Ledo Road (later renamed the Stilwell Road) from Ledo in Assam, India to Kunming in Yunnan, China, to enable Western aid to be sent to China; and airfields for the aircraft operating the Himalayan Air Bridge.

Grub pontoon bridge. IWM

135-metre-long Bailey suspension bridge over the River Shweli. https://www.thinkdefence.co.uk/2012/01/uk-military-bridging-equipment-the-bailey-bridge/

Up in the Air

In December 1943 the Third Tactical Air Force was formed to provide close air support for Army operations in India and Burma. Early actions included the low-level strafing and bombing of the besieging Japanese troops during the Battles of Kohima and Imphal.

Air drops by the RAF and USAAF were arranged for the Chindit deep penetration columns, and as the army advanced into Burma, landing strips were constructed by the RE from which the aircraft could operate.

The USAAF were actively operating 'The Hump', as the Himalayan Air Bridge was commonly known.

Among the RAF's ground engineers working round-the-clock to service the aircraft was Flight Lieutenant Charles Pringle who went on to become Air Marshal Sir Charles Pringle.

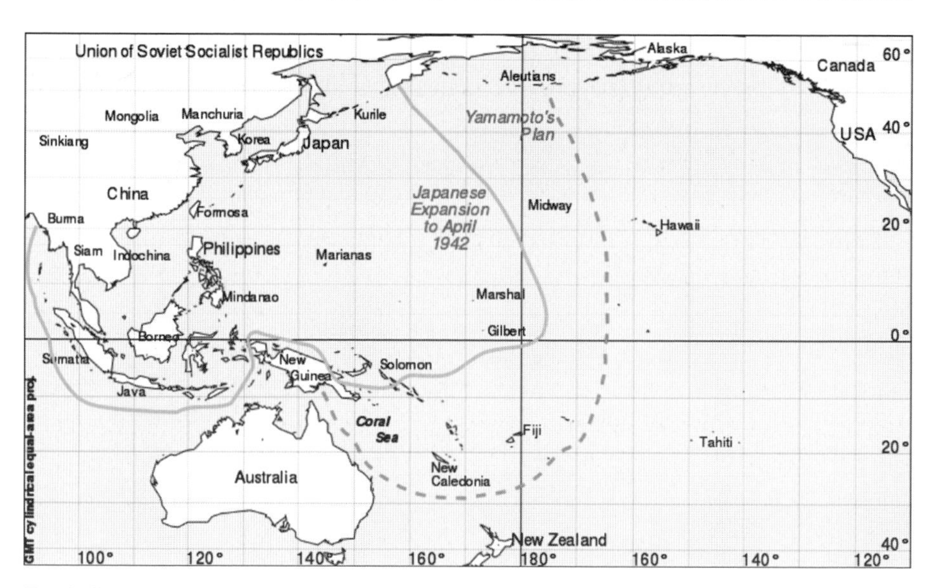

Pacific Ocean showing planned and actual Japanese conquests and key sites. Wikipedia

Fungus on the Wires

When South East Asia Command was established in August 1943, communication links were established from the headquarters in Kandy, Ceylon (now Sri Lanka), to various subsidiary headquarters, units, and ships.

Most of the Army's communication services in Southeast Asia were provided by the Indian Army. By the time of the Japanese attack on Malaya, it was severely depleted of trained personnel as the majority had already been deployed in the Middle East. Press communications were provided by a high-speed mobile transmitter at headquarters, where it was manned by Indian civilian personnel who sent bulletins to the commercial networks in Delhi, Calcutta (now Kolkata) and Bombay (now Mumbai). RSigs provided communication systems in ports and dockyards.

When the Japanese invasion of Burma needed to be repelled, communications were seriously hampered by the almost complete absence of telephones and roads in the border area as well as differing railway gauges. Furthermore, following the Japanese offensives they had dominance in the air and on sea, meaning the supply of equipment which had previously come from either the UK or Australia was now mostly not feasible.

Fighting was virtually impossible during the monsoon season, and at other times communications were seriously hampered by the high humidity which caused fungus and moisture on internal wiring, thus destroying circuit insulation. Drying equipment in improvised ovens mitigated the effect, but the specialised tropicalising of equipment during manufacture that was introduced for the later phases proved to be more effective.

Landlines were frequently damaged by operations to widen tracks, resulting in the creation of special working parties attached to road construction units to protect and repair the lines.

Some other communications features peculiar to the campaign in Southeast Asia were that three-quarters of the personnel were from the Indian Army; sets were provided in trucks when possible, although the poor road infrastructure meant that much of the equipment had to be carried by mules or the troops themselves; and wireless was used increasingly and, in certain situations (for example, Imphal), entirely.

In addition the large amount of jungle clearing that was needed slowed down the rate at which lines could be constructed, the damage a tropical storm could cause meant that the highest possible standard of construction was essential, the screening effect of trees greatly reduced the ground level range of wireless communications, the supply of stores was extremely difficult, air supply became a normal procedure, and generators made so much noise that they could disclose a unit's position and so were eventually not used at night.

Signal plans had to be based on an intimate cooperation between the three services (which resulted in the creation of a Combined Signal Board to handle mutual

Engineers in Command – and winning medals

Ex-sapper General L. Perowne CB CBE commanded a Chindit brigade, while one of the most successful Chindit columns was led by former RE officer Brigadier J. M. Calvert DSO.

The second of the RE's two Victoria Crosses was won by Lieutenant Claude Raymond RE who was in charge of a small unit on the Burmese coast which met and successfully fought a Japanese detachment, Raymond pressing home the attack despite being mortally wounded.

problems). It proved essential to have a comprehensive line system for command and administration which could cater for all three services as well as other users; the main methods of communication were mutually complementary with both line and wireless communication necessary at all levels.

REME and IEME Activities

In 1943, some months after the creation of REME, the Indian Army established the Indian Electrical and Mechanical Engineers (IEME) and the two corps undertook similar operations in Southeast Asia.

As other units discovered, the weather and humidity made both operations and equipment maintenance particularly difficult. The post-war review of REME operations stated that 'more equipment has been rendered unserviceable in storage than in actual operations'.

Stores and equipment arriving from overseas were generally disembarked at Bombay and then conveyed nearly 2,000 km across India to Calcutta and thence to various operational areas. This often involved a transfer from standard to metre gauge railway and a crossing by ferry of the 1.5 km-wide Brahmaputra River.

By their very nature the Chindit campaigns did not involve any wheeled transport, but it was essential that the forces' small arms and radios etc. should be fully functional at all times. Armourers formed part of each column, and forward workshops were built at each landing strip and stronghold. The demand for heavy weapons grew and eventually the ready-use stock contained items such as assault boats, small arms, mortars, flame throwers, PIATs, bazookas and other small-calibre artillery, wireless sets, batteries, and charging engines.

Some of the kit was new and occasionally arrived without instructions; in such cases the artificers first had to figure out how it worked and then instruct the troops who would be using it. When the Japanese started attacking the strongholds, the heavy equipment needed for their defence had to be stripped down and the parts marked and loaded onto gliders to be flown out to the bases, where REME and IEME personnel would reassemble them.

Indian Bletchleys

Following the success of the Bletchley Park codebreaking operations the British established a parallel organisation in India and achieved similar results decrypting Japanese signals. Many young people were sent on crash courses to learn Japanese so as to be able to participate.

American cryptographers, meanwhile, were able routinely to decipher Japanese signals through their Magic system, thereby giving advance warning of enemy positions and plans.

During the main advance into Burma in 1944–45 large numbers of air drops were used, and the need to prioritise the equipment to be carried became crucial. There was once no equipment capable of unloading a tank engine at a landing strip, and a fork-lift truck had to be flown in piecemeal and reassembled.

Naval Activities

As well as conventional naval activity in the Indian Ocean, naval forces were working in the relatively small-scale amphibious operations off the Burmese coast. After the successful Normandy invasion, landing craft and other vessels were moved to the Indian Ocean in preparation for an assault on the Malay Peninsula, while the Army undertook the waterproofing of vehicles and the training of Indian drivers, many of whom had not previously seen the sea.

On 26 July 1945 three XE midget submarines, manned by British, Australian, and New Zealand crews, successfully immobilised two Japanese heavy cruisers near Singapore. Concurrently, preparations for the invasion were nearing completion when the sudden Japanese surrender changed everything. But as it was uncertain whether the Japanese commander in Malaya would surrender without a fight, it was decided to proceed with the invasion and a large armada of ships carrying troops and vehicles

RN Fleet Air Arm Grumman Avengers, Supermarine Seafires, and Fairey Fireflies on the deck of HMS Implacable *warming up their engines before take-off. Collection Database of the Australian War Memorial under the ID Number:019037*

HMS Formidable *on fire after a kamikaze attack. IWM*

Lieutenant Yamaguchi's Yokosuka D4Y3 (Type 33) Suisei diving at Essex, 25 November 1944. The dive brakes are extended and the non-self-sealing port wing tank is trailing fuel vapour and/or smoke. US Navy Public domain

British Pacific Fleet

The British Pacific Fleet (BPF) was formed to fight alongside the Americans in the Pacific War. The US Naval Institute recorded in its *Naval History Magazine* (February 2013, Volume 27, Number 1) that, 'After more than five exhausting years of global conflict, the British Commonwealth organized a powerful modern fleet that fought as equal partners with the US Navy in the late stages of the Pacific war.'

Despite only being in existence from November 1944 until September 1948 it was the strongest fleet that the Royal Navy had ever assembled. It was commanded by Admiral Sir Bruce Fraser who had responsibilities to the Admiralty for the maintenance and welfare of the Fleet; to the Australian and New Zealand governments for shore-based support and activities; and to the US Admiral commanding the US Pacific Fleet for operational duties.

It had its main base in Australia, and at any one time comprised up to four battleships, six fleet carriers, four light carriers, nine escort carriers, and two maintenance carriers, together carrying over 750 aircraft. Also in the BPF were five cruisers, 11 destroyers, and several smaller vessels and submarines, a number of which (and the majority of aircrews), were from Commonwealth countries. While at sea, it was supported by a fleet train consisting of up to ten repair and maintenance ships, 22 tankers, 24 store carriers, four hospital ships, five tugs, two floating docks, and 11 miscellaneous vessels. It included HMS *Menestheus,* a minelayer which was converted to serve as an amenities vessel to provide recreational facilities for naval personnel between operations.

The BPF participated in actions against oil installations in Sumatra; the invasion of Okinawa (when its carriers' armoured flight decks proved more resilient to *kamikaze* attacks than the flight decks of the US carriers); submarine and midget submarine attacks on Japanese warships; and the bombardment of naval installations in the Japanese home islands.

was assembled off Port Dickson and Morib in western Malaya by 9 September 1945. Fortunately the subsequent landings were unopposed, although inexperience and inadequate training meant that a relatively large number of vehicles were drowned after leaving their landing craft. Many of these were the beach recovery BARVs, thus hampering the recovery of other vehicles. In the Morib landings 7,299 vehicles were landed, 824 were recovered from the sea, and 28 were lost.

As in Sicily and Normandy, the *sine qua non* for the success of the landings was the availability of appropriate landing craft for both men and vehicles. American engineers and industry designed and built these in the vast numbers that were needed in all the operational theatres.

The new generation of American submarines played a vital role in supporting the American advances across the Pacific by sinking Japanese warships and merchantmen and also by forward reconnaissance and guiding amphibious landing forces. They also rescued airmen whose planes were shot down, thereby raising the morale of air crews.

When the reformed BPF first joined the American Navy in 1944, the Royal Navy learned to its embarrassment that the vast distances of the Pacific meant that its ships could only stay on station for short periods before having to return to a base for refuelling. Conscious of similar problems, the Americans had developed the fleet train of tankers and supply ships, which, protected by destroyers, accompanied its various fleets, thereby enabling them to stay on operations for much longer periods – a technique that the British soon adopted.

The distances involved in the Pacific Ocean campaign also highlighted other deficiencies in the British vessels. In particular, the propulsion systems in many ships were simply not up to the distances they had to cover and the poundings they received on the high seas in the Pacific.

The Americans were also in the forefront of developing new aircraft carriers, a concept which had originally been pioneered by the RN during the First World War. 'The fast carrier,' said the US Navy's Admiral Sims in 1925, 'is the capital ship of the future.'

Interwar carriers were designed for the planes then available, but the development of faster and larger aircraft in the late 1930s meant that a new generation of carriers was required. The Americans responded by designing and building 24 Essex-class carriers, which were significantly larger, longer, and faster than their predecessors. The first of these ships was ordered in July 1940 and went into service in 1942. They also developed the Grumman F6F Hellcat fighter, their principal carrier-borne aircraft which saw notable service throughout the Pacific campaign.

USS Essex. *USN 36007 from the U.S. Naval Photographic Center*

As the Americans began their fightback against Japan, the two nations' navies fought each other in a number of important naval battles, with carrier-borne aircraft playing the dominant role in all of them. The first was the Battle of the Coral Sea. Fought on 4–8 May 1942, it was the first naval action ever in which aircraft carriers engaged each other, as well as the first in which neither side's ships sighted or fired directly upon the other. With both sides losing carriers and other ships, the result was a tactical victory for the Japanese but a strategic victory for the Americans and their Australian allies, as the planned Japanese assault on New Guinea was cancelled.

More decisive, however, was the Battle of Midway that took place on 3–7 June 1942, with the Japanese losing four heavy carriers and the Americans only one. Thereafter the Americans were not seriously challenged again, and the Battle of the Philippine Sea, fought on 19–20 June 1944, saw the Japanese carrier force virtually eliminated. Three of her carriers were lost, two of which were torpedoed by submarines, against one damaged American battleship. Among the factors which contributed greatly to the American victory were new anti-aircraft proximity fuses, improved tactics, better

The Seabees

Ben Moreell was a civil engineering graduate who had joined the US Navy during the First World War, achieved distinction in masterminding the development and construction of military bases, and caught the attention of future-president Franklin Roosevelt, who was then an assistant secretary of the Navy. Having been sent to Paris in the early 1930s to study at the prestigious *École Nationale Ponts et Chaussées*, Moreell was appointed by Roosevelt, now the US President, to be in charge of US Navy dockyards and chief civil engineer of the US Navy. Among his early projects was the construction of two dry docks at Pearl Harbor. Undamaged by the Japanese raid, these proved invaluable during the war.

Moreell recognised that in order to recover the Japanese-occupied Pacific Islands, it would be necessary to construct bases from which the American forces could operate, and he proposed to Roosevelt that naval construction battalions should be established. This was agreed to, and in March 1942 the US Navy's Construction Battalions were formally established. More generally known as Seabees or CBs, and with the motto *Construimus, Batuimus* (We Build, We Fight), their members were recruited from the American construction industry. By the war's end some 325,000 men had served with the Seabees.

While the Seabees also participated in the Mediterranean and Normandy landings, their major involvement was in the Pacific theatre, where they constructed 111 major airstrips, 441 piers, dockyards, ship and submarine bases, oil storage tanks, hospitals, recreational bases and housing for the troops. When necessary they fought alongside the Marines. Most vital of all were the five major airbases they built in the Marianas – no fewer than three on Tinian – all of which could accommodate the new generation of heavy and long-range Boeing B-29 super fortresses.

training of US pilots, and advance knowledge of Japanese positions obtained through signal interception and decoding.

Air Operations

Air cover for amphibious operations in south-east Asia was provided initially by the Fleet Air Arm (thus necessitating communication links with RN ships), but with the RAF generally taking over when the bridgehead had been established. In the Pacific, once airfields had been constructed on recently conquered islands, the new generation of American long-range heavy bombers could be used, and they were soon a dominant factor. The Boeing B-17 Flying Fortress was the principal heavy bomber until May 1944 when the Boeing B-29 Super Fortress went into service. Not only was it heavier and with a longer range than its predecessors but it could also fly at altitudes where it could not be attacked by enemy fighters or anti-aircraft fire. However, because of its many novel features, there were several teething problems to overcome before it could go into service. It also required longer and heavier runways, and the new airfields built in Tinian and elsewhere in the Marianas were specifically designed to accommodate it.

Although the Japanese mainland was now within range of the B-29s the Marianas were not yet ready to accommodate them, and so in June 1944 B-29s were based in India and refuelled in China prior to attacking targets in Japan. Logistically, however, this was a difficult operation for the Americans to mount and, although such raids continued over the following months, they could not become the main thrust at the

Super Fortresses at Tinian airfield. US Public domain

Japanese mainland. Then, on 9–10 March 1945 more than 300 Super Fortresses flew from the Marianas to attack Tokyo and caused devastation in the city. In the following days other Japanese cities were attacked. It has been calculated that altogether over 100 km^2 of industrial sites were devastated, while two million buildings were destroyed, and 13 million civilians lost their homes.

Raids continued over the next few months and then, on 6 and 9 August 1945, Hiroshima and Nagasaki respectively were destroyed by the first atomic bombs.

The A-bomb

The ultimate engineering impact on the war (taking the adjective literally) was the dropping of atomic bombs over Hiroshima and Nagasaki. In the early years of the 20th century various European and American scientists had undertaken basic research into the composition of the atom, and a few of them speculated that the energy contained in an atom could be used as an explosive device. While many leading scientists were sceptical, the concept was thought to be realistic by others. Authors such as H. G. Wells accepted the possibility and wrote about the concept, while as early as 1924 Winston Churchill wondered whether 'a bomb no bigger than an orange had a secret power to devastate a township at a stroke'.

In 1939–40 most leading scientists working in Britain were involved in the research and development of radar; those who were not included Otto Frisch and Rudolph Peierls, refugees from Austria and Germany respectively. They were working at the University of Birmingham and, as enemy aliens, had not been allowed to work on top-secret projects such as radar. In March 1940 they jointly wrote a paper entitled 'On the construction of a Super Bomb based on a nuclear chain reaction in Uranium'.

Chadwick, Sir James CH FRS (1891–1974)

A British physicist, he studied at the University of Manchester under Ernest Rutherford, 'the father of nuclear physics', and in Berlin under Hans Geiger, the inventor of the Geiger counter. While at the University of Liverpool Chadwick was awarded a Nobel Prize for the discovery of the neutron in 1935. During the Second World War he was a member of the British scientific committee charged with considering the feasibility of making an atomic bomb and was the author of the report given to the Americans which inspired them to undertake serious research efforts. Following the establishment of the Manhattan Project, Chadwick led the British team working with the Americans on the development of the bomb. After the war he became master of Gonville and Caius College, Cambridge, and retired in 1959.

After this was brought to the attention of Sir Henry Tizard (see Chapter 1), a committee of leading scientists was formed under the chairmanship of Professor G. P. Thomson of Imperial College, and an organisation with the cover name of Tube Alloys was created to progress the project. Among the committee members were Patrick Blackett (see Chapter 5) and James Chadwick. University research continued under Professor Mark Oliphant at Birmingham and Chadwick at Liverpool.

Among the committee's conclusions were: that it would be possible to make an effective bomb which – containing a mere 10 kg of active material – would have the destructive effect of some 1,800 tons of TNT and (on Blackett's strong recommendation) that it would be advantageous for the bomb to be developed in the USA.

Bush, Vannevar (1890–1974)

An American engineer, inventor, and science administrator, who throughout the Second World War headed the US Office of Scientific Research and Development (OSRD), which carried out almost all wartime military R&D. He had joined the Department of Electrical Engineering at Massachusetts Institute of Technology (MIT) in 1919, and founded the company now known as Raytheon in 1922. Starting in 1927, Bush constructed a differential analyser – an analog computer with some digital components that could solve differential equations with as many as 18 independent variables. An offshoot of the work at MIT by Bush and others was the beginning of digital circuit design theory. In 1932 Bush was appointed vice-president of MIT and dean of the MIT School of Engineering, and president of the Carnegie Institution of Washington in 1938. He was appointed to the National Advisory Committee for Aeronautics (NACA) in the same year and soon became chairman. As chairman of the National Defense Research Committee (NDRC), and later director of OSRD, Bush coordinated the activities of some 6,000 leading American scientists in the applications of science to warfare. As head of NDRC and OSRD, he initiated the Manhattan Project (to develop an atomic bomb) and ensured that it received top priority from the highest levels of government. In 'Science, The Endless Frontier', his 1945 report to the President of the United States, Bush called for an expansion of Government support for science, and he pressed for the creation of the National Science Foundation. He was awarded the Medal of Merit by President Truman in 1948, the National Medal of Science by President Lyndon Johnson in 1963, and the Atomic Pioneers Award by President Richard Nixon in February 1970. He was also made a Knight Commander of the Order of the British Empire in 1948, and an Officer of the French Legion of Honor in 1955. After suffering a stroke, he died in Belmont, Massachusetts, in 1974 aged 84.

Conant, James (1893–1978)

Conant was an American chemist who was a transformative president of Harvard University. After obtaining a PhD in chemistry from Harvard in 1916, he served with the US Army in the First World War working on the development of poison gases. Post-war he became a professor of chemistry at Harvard before being appointed president in 1933. During the Second World War he was chairman of the National Defence Research Committee, where he oversaw American input into the development of the atomic bomb. In 1953 he retired from Harvard and shortly afterwards was made the first US ambassador to West Germany. After returning to the USA in 1957 he suffered ill-health and wrote books on the American education system.

Groves, Lieutenant General Leslie (1896–1970)

The son of a United States Army chaplain, he was a US Corps of Engineers officer who oversaw the construction of the Pentagon prior to directing the Manhattan Project. He retired from the Army in 1948 and became a vice-president at Sperry Rand.

Meanwhile, although some low key research was proceeding in the United States, the concept was not being actively pursued. Even when details of the British work were passed to the top American scientists advising the government – including Vannevar Bush of MIT and James Conant, president of Harvard and chairman of the National Defence Research Committee – it was not acted upon for some time. On 9 October 1941, however, President Roosevelt authorised that work on the subject be undertaken at Columbia and Princeton universities.

Then, early in 1942, the Manhattan Project was authorised in which the Americans, with British and Canadian support, undertook the research and development work needed to create an atomic bomb. US Brigadier Leslie Groves was in charge, while others working on the project included the American Ernest Lawrence, the British John Cockcroft, and the German-born refugee Klaus Fuchs, who, post-war, was discovered to have been a Soviet spy.

Although some work was undertaken in Tennessee and Washington most of it was at Los Alamos, New Mexico, where a trial bomb was successfully detonated in

Lawrence, Ernest (1901–58)

Lawrence was a pioneering American nuclear scientist who won the Nobel Prize in Physics in 1939 for his invention of the cyclotron. Throughout the Second World War he worked on uranium-isotope separation for the Manhattan Project. A graduate of the University of South Dakota he also studied at the University of Minnesota and Yale University before becoming the youngest professor at the University of California in 1930. Following the war he campaigned for government sponsorship of large scientific programmes. After his death laboratories were named after him and chemical element number 103 was named 'Lawrencium' in his honour after its discovery at Berkeley in 1961.

Cockcroft, Sir John OM KCB CBE FRS (1897–1967)

A British physicist, he shared the Nobel Prize in physics for splitting the atomic nucleus with Ernest Walton, and was instrumental in the development of nuclear power. After graduating in mathematics at Manchester he served as a signaller in the First World War before obtaining further degrees in electrical engineering and mathematics from Manchester and Cambridge. In 1928 he began his work on splitting the atom. During the Second World War he worked on radar and the atomic bomb prior to leading the Canadian Atomic Energy Project. Returning to Britain in 1946 he set up the Atomic Energy Research Establishment and became its first director. He was the first master of Churchill College, Cambridge.

July 1945. By the end of the month President Truman, having succeeded Roosevelt following his death in April, authorised the dropping of the bombs on Hiroshima and Nagasaki.

Following the devastation caused by the bombs, on 15 August 1945 the Japanese Emperor broadcast his agreement of his country's unconditional surrender. A formal instrument of surrender was signed by leading Japanese dignitaries and military representatives of the United States, Great Britain, and other Allied countries on board the USS *Missouri* in Tokyo harbour on 2 September 1945. Similar ceremonies were also held in Hong Kong, Singapore, and other major cities.

While in recent years there has been some condemnation of the decision to drop the atomic bombs, little has been said about the many Allied and Japanese lives that

Fuchs, Klaus (1911–88)

A nuclear physicist and convicted spy, Fuchs was the son of a Lutheran pastor. He was born in Hesse, Germany, and studied at Leipzig where he got involved in student politics and joined the Communist Party. After the rise of the Nazi Party he fled to Britain, where he studied at the universities of Bristol and Edinburgh and acquired British nationality. On the outbreak of the Second World War he was interned in the Isle of Man and Canada but returned to the UK in 1941. He was allowed to work with his fellow German, Rudolf Peierls, on the atomic bomb project although shortly afterwards he started passing details of his work to a contact in the Soviet Embassy. In 1943 Fuchs and Peierls went to the USA to work on the Manhattan Project. After the war he returned to the UK and worked at the Atomic Energy Research Establishment at Harwell. In January 1950 he confessed that he was a spy and was sentenced to 14 years' imprisonment and stripped of his British citizenship. He was released in 1959 and emigrated to East Germany where he was elected to the Academy of Sciences and the central committee of Socialist Unity Party of Germany. He was later appointed deputy director of the Institute for Nuclear Research in Rossendorf, where he served until he retired in 1979.

Hiroshima 6 August 1945

Nagasaki 9 August 1945

were saved because the swift Japanese surrender precluded an invasion of the Japanese mainland that would almost certainly have resulted in protracted and bloody fighting over many months.

There followed a poignant experience for one young engineer, Lieutenant (E) John Corney RN, an officer in HMS *Sussex* which was one of the first ships to return to Singapore. When the Japanese invaded Malaya Corney was at school in the UK while his father, who had been working in the country, was incarcerated at the notorious Changi jail and had had no news of his family ever since. Shortly after *Sussex* went alongside in Singapore, young Corney and the ship's chaplain were allowed to go ashore to search respectively for his father and the Bishop of Singapore. One can only imagine the emotions of father and son at their reunion.

Peace was now restored around the world. People could reflect on the years of conflict, begin reconciliation between former enemies, and start the reconstruction of their ravaged countries...

Chapter 11

THE 3RS OF THE POST-WAR WORLD

Lest we forget

The three Rs of the post-world war are:

Remembrance
Reconciliation
Reconstruction

Remembrance

The principal act of remembrance today is gratitude to all those who served or were involved in the Second World War and, in particular, to the 20 million combatants and civilians who lost their lives. We should never forget the men and women whose dedication, bravery, and downright doggedness ensured that the worst never happened. The front line heroes and those who fought off the German air raids over southern England have been fittingly anthologised, and the Churchill-inspired '*Blitz* spirit' is an abiding feature of histories of the Home Front.

But it is also appropriate to remember the various critical actions and developments which made a significant contribution to ultimate victory, many of which have been described in this book.

It is also fitting to pay tribute to the ingenuity shown by the many Allied prisoners of war (POWs) who tried to escape from captivity and who conceived and built remarkable devices in their attempts. In addition, it is salutary to recall that the enemy also made a number of significant technological advances. These are described in Appendices 7 and 8 respectively.

Reconciliation

Trade has been one of the most effective means of restoring amicable relations. It began soon after the Second World War with the establishment of the European Coal and Steel Community (ECSC) in 1951. The ECSC was first proposed by French Foreign Minister Robert Schuman to 'make war not only unthinkable but materially impossible' and was officially established in 1951 by the Treaty of Paris. Signed by Belgium, France, Italy, Luxembourg, the Netherlands, and West Germany it started the formal process of integration between European nations which has ultimately led to the creation of the European Union.

At an individual level there was initially some reluctance to purchase cars made by former enemies, let alone travel to those countries, but over time considerations of quality, value for money, and interest overrode such objections. Today both German and Japanese cars are among the most popular in the UK, while many companies have established manufacturing plants in these countries. The engineers responsible for the construction of the factories and for the design and manufacture of the vehicles have played major roles in all these operations.

Twinning between towns and villages of the former warring states has been a popular and frequent form of reconciliation. A prime example is the connection between Coventry and Dresden, two of the most heavily bombed cities of the conflict.

Furthermore, a number of people, often some of those who had been ill treated as prisoners of war, have in the fullness of time returned to the country where they were interned with the aim of finding and forgiving their former captors. Among these was the railway buff and engineer Eric Lomax who, as a young Royal Signals officer, was taken prisoner at Singapore and made to work on the infamous Burma railway. After a clandestine radio he had made was discovered he was tortured and severely dealt with.

Following liberation at the end of the war he returned to civilian life and in his retirement learned that his principal torturer had repented of his actions, written a book, and built a temple at the site of the railway. Lomax then went to Thailand to meet him, and the two became reconciled. He then wrote of his experiences in an autobiography entitled *The Railway Man*, published in 1995, which has been described as 'an extraordinary story of torture and reconciliation'. A television adaptation was made, and a film with the same title was released in 2013 starring Colin Firth as Lomax and Nicole Kidman as his second wife. He died in 2012 aged 93.

Reconstruction

All the major cities in the warring countries suffered significant destruction of buildings and infrastructure, and one of the immediate post-war tasks was their reconstruction. It is remarkable how quickly and effectively this was often achieved. The before and after photographs on the following pages compare post-war destruction with reconstructed sites; these are but two examples of similar scenes across much of Europe and Asia. Engineers, together with professionals in associated disciplines, have played huge roles in both the planning and execution of the reconstruction of cities, buildings, and infrastructure. Some buildings have been rebuilt in their original style.

The Frauenkirche, Dresden in 1945 (upper) and in 1991 (lower) after reconstruction. Bundesarchiv, B 145 Bild-F088675-0031 / Faßbender, Julia / CC-BY-SA 3.0

The gutted Hiroshima Industrial Promotion Hall (now the A-Bomb Dome).

The same view today

What Might Have Been

Since this book concentrates on the technical achievements involved, it is worthwhile contemplating what would have happened had they not been developed or applied, particularly in the vital early battles.

If Britain had not won the Battle of Britain, Germany would have been able to continue its aerial bombardment of the UK – Britain would almost certainly have had to negotiate a settlement or surrender. Germany would then have been able to concentrate all her resources on her invasion of Russia and would almost certainly have been victorious there. As a result Nazi Germany would have become the undisputed master of Europe.

If, despite having won the Battle of Britain, the UK had then lost the Battle of the Atlantic, her vital trans-Atlantic lifeline conveying armaments, fuel, food, and other essential supplies would have been cut and she would have been unable to continue the war. Furthermore, there could then have been no American participation in North Africa or in Europe, and Britain would not have been able to participate in any activity in Southeast Asia or the Pacific. While the US, thanks to her greater resources of technology and manpower, would almost certainly have eventually emerged victorious in that theatre of war without Allied support, the political landscape in the Pacific, Southeast Asia, and Australasia would almost certainly have been very different to what it has been over the past half a century.

It is also possible that, had the US been the only power to possess atomic weapons, and had Germany provoked her, might some German cities have been laid waste in the same way as Hiroshima and Nagasaki?

Who can confidently say that victory would have been achieved without the ability to track enemy air attacks with the use of radar? Or without the Spitfires and Hurricanes that won the Battle of Britain? Or without the Merlin engines which powered them? Or without the breaking of the German codes at Bletchley Park? Or without the new technology that finally beat the U-boat menace? Or without 'Hobo's Funnies' that first breached the Atlantic Wall? Or without the Mulberry harbour that enabled the Normandy bridgehead to be sustained and reinforced? Or without the new generation of aircraft carriers and planes that defeated the Japanese Navy in the Pacific? Or without the brains and technology that conceived and constructed the atomic bomb?

This book has attempted to do justice to those hitherto largely unsung scientists and engineers whose brains backed the valour; and sheer hard labour of those serving under arms or working in factories; those who maintained near-normal services in appalling conditions; and the men and women who aided the war effort in countless other ways.

War, it is rightly said, always encourages technical developments, and never was greater ingenuity shown than by the scientists and engineers engaged in conceiving, developing and producing the novel machines, weapons and structures that made victory possible in the Second World War.

It is surely no exaggeration to say that their role and achievements were vital in overcoming the threats of the Axis powers, putting us all forever in their debt.

Appendix 1

The Engineers who Thought and Fought

The new systems, weapons and machines that played such a vital part in achieving ultimate victory were conceived and developed by a small number of – mostly civilian – scientists and engineers. Tens of thousands of other civilians manufactured the equipment which in turn was used by hundreds of thousands of service personnel, many of whom were in the technical branches of the fighting services.

Vast numbers of engineers and craftsmen served in the armed forces and many more civilians were trained for service in the technical branches of the armed services. Lots of officers were qualified to degree-level standards while non-commissioned officers were proficient in their particular craft or trade. The principal technical branches of the three services are described here.

Royal Navy (RN)

Its five main technical branches were:

1. Marine engineering – Looked after a ship or submarine's propulsion and other machinery and led damage-control operations
2. Electrical engineering – Oversaw a vessel's electrical systems
3. Special branch – In charge of radar and other electronic equipment
4. Ordnance engineering – Responsible for maintaining a ship's armaments
5. Air engineering – Serviced the aircraft of the Fleet Air Arm

A few engineers also served with the Royal Marines (RM) who undertook commando-style amphibious operations; some RM were also stationed in larger ships.

Army

The Army had three main technical branches:

1. Royal Engineers (RE) – Led assaults on beaches; cleared and laid mines; built and destroyed bridges and roads; operated railways, inland transport and postal services; surveyed terrain and produced maps; supplied water and fuel; and built and maintained Army establishments

2. Royal Signals (RSigs) – Provided, maintained, and operated radio and telephonic communications
3. Royal Electrical and Mechanical Engineers (REME) – Recovered and repaired damaged vehicles and ordnance

As the latter two corps both had their origins in the Royal Engineers, Field-Marshal Montgomery's post-war tribute to the Sappers aptly summarises the wartime achievements of all three corps:

> *The Sappers really need no tribute from me; their reward lies in the glory of their achievement. The more science intervenes in warfare, the more will be the need for engineers in field armies; in the late war there were never enough Sappers at any time. Their special tasks involved the upkeep and repair of communications; roads, bridges, railways, canals, mine sweeping. The Sappers rose to great heights in World War II and their contribution to victory was beyond all calculations.*

Royal Air Force (RAF)

The original vision of Lord Trenchard, the 'father' of the RAF, was that all RAF officers would be pilots other than a few specialists such as medical and accounting officers. However, things soon changed. Although there had been technical officers in the RFC and RNAS, for much of the inter-war period officers of the General Duties Branch (pilots) were selected to undertake specialist training in subjects such as engineering, armament, navigation, and photography. In 1940 the decision was taken to establish a technical branch in both the regular and non-regular air forces for officers employed in engineering, signals, and armament duties.

Officers in the Technical Branch were responsible for servicing and maintaining the aircraft. They also supervised the work of non-commissioned officers (NCOs) and airmen in flying squadrons and non-flying units, and worked at headquarters and in the Air Ministry to manage projects. A number of engineers flew in bombers while others built landing strips and temporary airfields. Some would have been involved in developing new technologies or investigating enemy aircraft, whether by examining crashed and captured examples or exploiting other forms of intelligence.

Other engineers commissioned into the Branch during the war worked in roles related to their civilian work. For example, Group Captain Conrad Verity was an electrical engineer who worked in an Air Intelligence branch building up information on potential targets in the German electrical system. This included hydro-electric power from dams, and he was instrumental in collating target information for the 'Dam Busters' raid of May 1943.

The increasing number of squadrons and complexity of aircraft (in addition to new technologies such as radar) brought a need for more ground crews to support them. Trenchard's pre-war apprentice schemes had been designed to create skilled tradesmen who would be able to lead and manage less skilled personnel; their skills and ingenuity proved priceless on many occasions.

Other Engineers

Merchant Navy Engineers

The men of the Merchant Navy (MN) were responsible for supplying Britain with the majority of the oil, food, weapons, and raw materials needed to sustain the country's ability to fight. Despite being classified as civilians, they suffered horrendously high casualty rates and were in many ways the unsung heroes of the war.

Even more exposed than their RN counterparts, MN engineers operated in unarmed vessels that travelled in convoys at the speed of the slowest vessel and which were mostly driven by reciprocating steam engines. Making smoke was a cardinal sin for these vessels as it could be seen over the horizon and disclose a convoy's position to U-boats or searching aircraft.

Civilian Engineers

Civilian engineers supported the fighting services, fought on the home front or contributed in countless other ways. Principal among them were:

- The scientists and engineers working in universities, research establishments, laboratories, and factories. They conceived, invented, tested and developed new technology before it was produced and made available for use.
- The engineers who, during the *Blitz* and the 1944 onslaughts by the V1 flying bombs and V2 rockets, maintained and reinstated Britain's bomb-damaged transport, communications and fire-fighting infrastructure. Thus they ensured the country could continue to resist and fight on with a stable morale.
- The countless millions who worked in offices and factories to design and produce vital equipment, plants and munitions. Often they continued undaunted after their homes or factories had suffered bomb damage.
- The construction workers who created and built the new military establishments, airfields, factories, Mulberry harbours and their associated infrastructure systems.

Research and Development Scientists and Engineers

Research and development technologists mostly worked for departments established by the Admiralty, War Office and Air Ministry. They also worked in academia and for the companies that manufactured the engines, planes, ships, tanks, munitions and electronic and other equipment used by the fighting services.

Others served on various scientific committees established by government departments. In addition, there were a few freelance inventors, including Ronald Hamilton who conceived the Swiss Roll floating roadways and the Lily floating islands. (See appendix 4).

Involvement of Academics

Many top academics were recruited directly by various government departments. Best known are those who worked at Bletchley Park although several others worked elsewhere on a range of duties.

German Jewish Scientists

Following the Nazi persecution of the Jews in the inter-war period, lots of scientists and engineers emigrated to Britain, France, and the USA. Thanks to the international links that existed, particularly between academic scientists, a number joined British universities. A leading British scientist kept a model of Adolf Hitler on his desk to remind him of the number of refugees who were now working with him.

Equipment Designers and Manufacturers

In Britain aircraft were designed and built by civilian companies, generally to Air Ministry specifications. Warships were designed by the Royal Corps of Naval Constructors and built in the Royal Dockyards or by civilian shipbuilding companies. Merchantmen were designed and built by the latter. Most weapons were manufactured at the Royal Arsenals at Woolwich and Enfield or the newly constructed Royal Ordnance Factories, while tanks and most specialist equipment were built in civilian factories.

Structures were largely designed by the Ministry of Works alongside the RE or by civilian consultants and built by the RE or civilian companies.

Reserved Occupations and State Bursaries

It was recognised that many of the technical roles vital to prosecuting the war and supporting the fighting services would have to be undertaken by civilian scientists and engineers. Such positions were classified as 'reserved occupations' which exempted the holders from conscription into the armed services. In some cases persons with appropriate qualifications were drafted into such positions.

Some 50,000 young persons (the author amongst them) were awarded state bursaries to enable them to study technological subjects prior to going into the fighting services, industry or academia.

Other Civilians Involved in the War Effort

Supporting both the military and civilian populations were:

- The coal miners and the gas and electricity generating staff who provided the country's necessary energy

- The dockers and the road and rail workers who transported imports and home-produced goods across the country
- The nutritionists who determined the national diet
- The farmers, gardeners and allotment holders who produced home-grown food

Technical Books and Papers

Most of the books written about the history of the Second World War make reference to some of the technical equipment used and there are a small number of books which have aimed to deal specifically with the subject. Many are listed in the Bibliography.

Additionally, most of the professional engineering institutions and societies have published papers describing some specific aspect of their members' involvement. For instance, there were 68 papers presented and discussed at the conference entitled *The Civil Engineer at War* held at the Institution of Civil Engineers (ICE) in 1948.

Appendix 2

Technology Applied to Weaponry

Victory in battle has always resulted from a combination of strategy, tactics, leadership, troop numbers, bravery, and luck – and the effective use of armaments that embraced the latest technology. The Second World War was the supreme example of this last aspect. The inventions that had a critical effect on the conduct and outcome of the war fell into three categories:

1. Those which were vital in achieving victory in the battles that ensured eventual Allied success, like radar, the Spitfire and Hurricane fighters, the inventions which defeated the U-boats in the Battle of the Atlantic, and the atomic bomb
2. Those which had a major effect on the course of the war e.g. the Bletchley Park decrypts, Bailey bridges, and the Mulberry harbours
3. Those whose impact was less but still made an important contribution to the overall outcome, for example the pipelines under the ocean, camouflage, deceptions and the wartime diet

Factors Contributing to the Development of Inventions

Among the factors responsible for the development of new ideas and their successful application were:

- Pressure of events such as the shipping losses caused by magnetic mines
- Response to enemy innovations like the German U-boat campaign and their *Blitzkrieg* warfare which necessitated close liaison between the Army and Air Force
- Foresight, e.g. the development and application of radar
- Organisation and the human factor – the arrangements for ensuring technologists were known to the authorities and were engaged in appropriate tasks and that there was effective dialogue between the military and technologists, as exemplified by the Sunday Soviets established by the Telecommunications Research Establishment at Malvern.

Typical Developmental Phases for Inventions

In a paper on the development of radar P. E. Judkins listed the following developmental phases:

- Conception – recognition of a need and an idea as to how it might be solved
- Financial support by a higher authority for the various stages of its development
- Experimentation and development – to research the concept and turn it into a practical application
- Confidence that it will perform e.g. RAF Fighter Command's recognition that radar would enable it to decide when to get fighters airborne and where they would be needed
- Acceptance by senior staff
- Logistics – the manufacture of the equipment in the numbers needed, the provision of spares, and ensuring all are provided where they will be needed
- Personnel – the selection, training, and organisation of persons to operate and maintain the equipment
- Information on how the equipment is to be used
- Training of operatives and feedback so as to improve subsequent models
- Interoperability to ensure it can be used in conjunction with other equipment

Factors Impeding New Developments

Guy Hartcup listed four reasons why developments might not proceed quickly enough, or at all:

1. Perceived lack of requirement, for instance, in the early 1930s neither the Admiralty nor the War Office considered radar to be important and did not pursue it until the RAF demonstrated its potential
2. Inter-service rivalry, e.g. responsibility for the Mulberry harbours was disputed between the Admiralty and the War Office
3. Excessive secrecy or caution which could limit the value of a new weapon – both proximity fuses and the Hedgehog depth charge launcher could have been used more effectively if their properties had been more widely known
4. The pace of development which could vary from weeks to years

Countermeasures

Much effort also went into creating countermeasures to nullify the effectiveness of the enemy's attacks. Both sides – the Germans more than the Allies – sometimes adopted a 'if not invented here, could not have been invented there' attitude. The Germans were guilty of this during the Battle of Britain when trying to decide how British radar operated while the British, during the V2 attacks on London, initially could not believe that the Germans had mastered the science of rocketry.

Appendix 3

MD1 'Winston Churchill's Workshop'

Among the weapons and devices invented by MD1 were:

- Limpet mines designed to be placed by frogmen on ships' hulls to which they became attached by magnetic attraction and later detonated by a time-delayed fuse. They brought Macrae to the attention of Jefferis, who had read about a particularly strong magnet in Macrae's magazine. Macrae was perplexed as to how to design a trigger that would activate a release mechanism following a period under water; he experimented with aniseed balls after noticing a colleague's children sucking them. These proved to be ideal and were used in production mines until a more scientific alternative could be devised. It was the department's first success – over half a million limpet mines were manufactured and caused many enemy ships to be sunk. The Royal Commission on Awards to Inventors rewarded the designers.
- Railway sabotage units (light camouflet sets) containing explosives. Placed under railway lines, they were intended to go off when activated by the deflections caused by a passing train, the ingenious part being the pressure switch which activated the mechanisms. Also adopted for other devices, over 2 million switches were made and issued during the course of the war.
- Floating river mines (W-bombs) created to cause problems for German river traffic on the Rhine and other Continental rivers. Conceived by Churchill while he was with the Admiralty, MDI needed to design mechanisms which allowed the explosives to lie undetected on the river bottom and then rise to just below the surface; they would then be carried by the current until they made contact with a barge or bridge abutment and then detonate. Models were made and successfully demonstrated by Jefferis and Macrae to the Cabinet and the French High Command. But with the Phoney War still on, the project was vetoed by the French for fear of inviting reprisals by the Germans. After the German invasion of France, however, nearly 2,000 of these devices, which became known as W-bombs, were released into the Rhine and caused immediate and major disruption to river traffic, although the German occupation of eastern France made further discharges impossible. They were used successfully elsewhere though and some 20,000 were manufactured and deployed, mostly by the

RAF. Yet the most important outcome was that Churchill became aware of the capabilities and potential of Jefferis and Macrae and their associates.

- Booby trap switches (activators) designed by Jefferis and Macrae: the Pressure Switch, used to sabotage railway lines; the Pull Switch which was based on an old-fashioned collar stud and had a moveable central spike; the Universal Switch that superseded both; and the Anti-Personnel (AP) Switch
- The Blacker Bombard (or 29 mm spigot mortar) – a mortar-type missile conceived by Lieutenant-Colonel Blacker. Successfully demonstrated to Churchill and other VIPs at Chequers in the summer of 1940, it was then issued to the Home Guard as a matter of urgency. The Royal Commission awarded Blacker £7,000 for this design (as well as £25,000 for other inventions (awards 142–147)).
- The PIAT (Projector Infantry Anti-Tank) weapon, a spigot mortar developed from Lieutenant-Colonel Blacker's concepts which fired a 1.1 kg bomb. It was one of the most successful short-range anti-tank weapons of the Second World War and was much used by the infantry. Six Victoria Crosses were reputedly won by soldiers using them.
- Sticky Bombs (later named ST grenades) created to be thrown at tanks to which they would adhere before exploding. After seeing a successful demonstration, the Prime Minister wrote one of his shortest memos on 2 October 1940: 'Sticky Bomb. Make one million. WSC.' They were also issued to the Home Guard and an award of £1,850 made for the invention of the sticky material used, (awards 3 and 4) but no award was given for the design of the bomb itself.
- The Time Pencil, a slim anti-personnel mine that went through many iterations before its design was finalised
- The Aero Switch, a small bomb which saboteurs placed in the wings of aircraft; they exploded when the air pressure dropped to a certain level
- The 'M' (or Macrae) Mine, another anti-personnel device
- Anti-lift devices incorporated in mines so that they would explode if the enemy tried to lift them
- Acoustic decoys that generated a sound similar to a ship's engines and were developed in conjunction with DMWD to detonate marine acoustic mines ahead of a ship
- Beehive hollow charges that were created in a variety of versions including CS for attacking capital ships and AS against submarines
- The JW (Johnnie Walker) Bomb, conceived to be dropped into harbours where it would seek out a ship and explode against it
- The Great Eastern bridge, developed by MD1 as a tank-mounted Bailey bridge which could rapidly enable an obstacle to be crossed

Appendix 4

Department of Miscellaneous Weapons Development (DMWD)

The Wheezers & Dodgers

Among the principal developments of the DMWD's technical activities were:

- Plastic armour (now known as composite armour) to reduce secondary damage and casualties resulting from splinters following a ship being hit. It was DMWD's first major success and was widely adopted, being applied to some 10,000 British and American ships. Lieutenant Commander E. Terrell RNVR was awarded £9,500 by the Royal Commission on Awards to Inventors while Neuchatel Asphalte Co. Ltd received £2,000 (awards 89 and 91). Dr Glanville of the Road Research Laboratory received nothing, probably because he was working in a government-funded research establishment.
- Vertical flame throwers installed on ships for attacking low flying aircraft – only partially successful
- Acoustic warning devices fitted on mastheads to give warning of approaching aircraft prior to the development of radar
- Dazzle guns to blind pilots in attacking planes. Ineffective and abandoned (but today being developed as directed energy weapons/lasers).
- Oerlikon anti-aircraft guns, developed in Switzerland in the mid-1930s and superior to any comparable gun. The Admiralty had accordingly placed orders, though following the fall of France it was not possible to import them. Fortunately, a former RN officer who had been inspecting their manufacture was able to carry copies of the designs and other critical information to Britain. He encountered many difficulties, however, and was forced to travel via Turkey, Palestine, and Egypt. Although it would not normally have been Goodeve's responsibility, his reputation for being able to 'get things done quickly' prompted the Admiralty to entrust him with arranging for the production of the gun at a new factory. Faced with many problems he eventually was able to arrange for the manufacture at a factory in Ruislip, Middlesex, which created 750–1,000 guns a month with half as many again being produced in the West Country. The US Navy also adopted the gun and installed as many as 90 in some ships.

- Rocket projectiles, conceived as an effective alternative to conventional guns but the initial tests proved disastrous and they had to be abandoned
- Rocket-propelled armour-penetrating spears designed to be launched from aircraft and to penetrate U-boat hulls. The device secured its first 'kill' within eight weeks of its introduction.
- Parachute and cable (PAC) devices, initially created by the RAF as a means of providing an instantaneous balloon barrage around a potential target. After heavy losses of merchant ships, however, it was decided to adapt them for the protection of vessels at sea. Initially installed in the bows of a ship, but later in the stern as well, the device was propelled vertically upwards by a rocket as a plane approached. Frequently successful in making the pilot abort the attack, they also resulted in the destruction of a number of enemy planes.
- Starshell, developed to enable the RN's Snowflake flare to be held in the sky long enough for convoy escort vessels to see a surfaced U-boat. This was achieved by attaching it to the head of a PAC rocket, effectively turning night into day. Commander F. D. Richardson RNVR was subsequently awarded £500 by the Royal Commission on Awards to Inventors.
- Holman Projector (potato thrower), a joint project between the DMWD and Holman manufacturers in Cornwall. A projectile was fired from a tube (working from the principle of a mortar), the motive force originally coming from compressed air but later by high pressure steam. Installed both as an offensive weapon on motor gun boats and an anti-aircraft weapon on steam-powered trawlers etc., some 4,500 were in service by 1943.
- Free balloon barrages which comprised a number of hydrogen-filled balloons supporting piano wires to which explosives were attached. The concept was abandoned after disastrous trials.
- Hedgehog anti-submarine depth charge launchers, although these were not part of DMWD's original remit of developing anti-aircraft devices. Developed jointly with Millis of MD1, they enabled an escort vessel to remain in Asdic contact with the submerged U-boat while the explosives detonated. Goodeve and Millis persisted in spite of disappointing trials and it became one of the most successful anti-submarine weapons of the war, being credited with some 50 U-boat kills.
- Assisting Barnes Wallis in testing variants of the Tallboy bomb
- The development and testing of radar counter-devices
- The supply of fresh water. Based on his experience with airships, Lt. Cdr N. S. Norway RNVR created a means of distilling sea or brackish water using the exhaust gases of internal combustion engines to produce drinkable water in powered lifeboats. The Army, who also needed fresh water in the Western Desert, supported the work, but by the time the technique was advanced the fighting had moved on.

- Model target planes for training anti-aircraft gunners in training establishments in the UK, Egypt and Canada. Wren personnel were involved in their manufacture and deployment.
- Successful results with camouflaging ships. Much less successful were attempts to conceal the Mersey Estuary and the Thames in Central and eastern London by spraying coal dust or soot over the water.
- Swiss Roll floating roadways, the brainchild of Ronald Hamilton, a brilliant inventor who was ahead of his time. His inventions generally failed to find a sponsor or result in benefits for Hamilton if one was found. Conducting his own experiments in a 60 metre tank in a bombed-out wing of the Grosvenor Hotel, Hamilton developed the floating roadway which could be rolled up for transport to the site – hence its name. Taken over by DMWD, it was used by the Royal Navy for landing its stores and personnel on the beach at Arromanches, Normandy. Hamilton was reputed to have been awarded £8,000 for his inventions.
- Lily floating islands, another invention by Hamilton, conceived as floating runways for aircraft. Again developed by DMWD, and after successful full scale trials in Scotland, they were produced to provide bases for air cover for vessels operating in the Far East. However, the dropping of the atomic bombs forestalled their use.
- The Great Panjandrum, a giant rolling mass of explosives propelled by rockets attached to its two large wheels. Designed to breach the Atlantic Wall, teething troubles with its steering mechanism meant it never saw action.
- The Alligator, an amphibious variant of the Great Panjandrum; although the steering problems were largely mastered, it also never saw action
- Perspex boats developed for landing agents and survey parties on enemy coasts
- Beach gradient meters were considered at the request of Combined Operations command; they needed them so as to be able to rapidly record beach contours on possible invasion beaches
- Rocket landing craft that enabled invasion forces to fire at enemy positions during the interval between the end of the naval bombardment and the actual landing. Becoming operational after some (literally) hair-scorching experimental work, they proved highly successful in both the Mediterranean and Normandy landings and were regarded by some as the most valuable of all DMWD's projects.
- The bubble water calmer, an attempt to produce calm conditions in the projected harbour to be built off the Normandy coast. Despite exhaustive experiments and calculations, it did not prove to be a practicable proposition, partly because of the power requirements.
- Hedgerow land-mine detonators were developed from the Hedgehog mine launcher, the essential links being the forward-launch concept and the fuse

needed to detonate the explosives at the right instant. Mounted on landing craft, they fired their charges as the craft approached the landing beaches so as to detonate any mines laid on the beach. Not a success in the Mediterranean, but a number proved their worth in Normandy.

- The Wreck Dispersal Pistol was intended to be a means of simultaneously detonating a number of explosive charges attached to underwater obstructions so as to clear harbours that had been blocked by the enemy. Lt Cdr A C Brimstead RNVR was awarded £1,500 by the Royal Commission on Awards to Inventors.
- Mulberry harbours, thought of and masterminded by others but Robert Lochner was the one who attended the meetings in Washington when the concept was discussed and approved by the heads of government and combined chiefs of staff. Thereafter DMWD were involved with testing and developing various aspects including:
    - Floating breakwaters to provide smooth water inside the Mulberry harbours. DMWD's Lochner led the development of the Lilo breakwater which was based on the principle of an inflated balloon while others conceived the rigid Bombardon floating breakwaters. After the decision was taken to adopt the Bombardon breakwaters their detailed design was tested and refined by DMWD in tanks in London and Haslar. Lochner, who realised that waves were confined to the surface and thus could be dampened by a floating obstacle, was awarded £1,000 by the Royal Commission on Awards to Inventors (award No. 19).
    - Kapok underwater jackets developed by the RN Medical Branch in conjunction with DMWD to protect frogmen working to defuse unexploded mines
    - Rocket-propelled grapnels which were used successfully by the Americans to scale the 30-metre-high coastal cliffs at Pointe du Hoc during the Omaha beach landings
    - Extendable ladders adapted from fire-escape ladders and mounted on landing craft that were also used to scale the Pointe du Hoc cliffs
    - Helter-skelter chute to speed the transfer of men from troopships to landing craft
    - Radar marking buoys to guide the invasion fleet at night
    - Radar decoys to mislead and confuse the enemy as to what was reflecting signals

Appendix 5

Anglo-American Air Offensive

The air campaign against German targets on the Continent grew from the small raids of 1939–40 by twin-engine RAF bombers to the major Anglo-American offensive of the last three years of the war. The RAF mounted night raids by fleets of over 1,000 bombers and the US Army Air Force (USAAF) concentrated on daytime raids, mostly by Boeing Super Fortresses and Consolidated Liberators and protected by long-range fighters. To make this possible nearly 450 new airfields were constructed in Britain and increasingly sophisticated technical aids were developed to improve bombing accuracy and enable returning aircraft to land in foggy conditions.

While raids were mostly on industrial targets and transport infrastructure, particular raids were also mounted on the German battleship *Tirpitz* and the dams in the Ruhr Valley, (special bombs were developed for both of these); on U-boat pens; and on the manufacturing and launch sites for the V1 flying bombs and V2 supersonic rockets that were targeted at Britain and the Low Countries in 1944–45.

Although the bombing campaign did not achieve the anticipated result of inducing the German surrender, it nevertheless crippled many aspects of the German war machine and diverted German resources from other fronts and activities. It also was the catalyst for a whole host of technical developments, many of which have made significant contributions to improving post-war life.

The high civilian casualties in Germany and the occupied countries and the virtual destruction of cities such as Dresden have caused many recent commentators to criticise aspects of the campaign. It is probably for such reasons that no British medal has been awarded for those who participated, despite the very high casualties suffered by British (and American) air crews – the attrition rate of trained RAF personnel was greater than that of junior officers in the Battle of the Somme in the First World War, while a young American had a better chance of survival fighting as a Marine in the Pacific than flying with the USAAF over Europe.

However, it is easy to criticise with hindsight and for today's commentators not fully to appreciate the circumstances of the time and the pressures operating on the decision makers of the day. The recently erected memorial to Bomber Command near Hyde Park Corner and the American Cemetery and Memorial at Madingley,

Cambridge, are poignant tributes to those who participated and lost their lives in the campaign.

Airfield Construction

Before 1939 most RAF airfields were on level areas of grass which drained quickly. With increasing weights of aircraft it was evident that concrete or tarmacadam-paved runways were needed. By 1942 the specification for a standard operational airfield was a 1,250-metre-long paved strip running approximately northeast to southwest and two subsidiary paved strips each some 900 metres long at approximately 60° to the main runway. Longer strips were provided at airfields for bombers and heavy transport planes. Dispersal points, hard standings, and airfield buildings were also built. The principal additional design considerations included ground consolidation, concrete and tarmacadam specifications, expansion joints, and runway drainage.

During the six years of the war 444 new airfields were constructed in the UK and in the peak year of 1942 new airfields were being commissioned at an average rate of one every three days. There were also 63 major extensions to existing stations.

Working under Air Ministry direction, the constructional work was mostly carried out by civilian contractors, many employing large numbers of Irish labourers and other operatives (even though Ireland was neutral during the war). Altogether some 136 separate contractors were involved employing at their peak over 130,000 men. Fifty of the new airfields were used by the USAAF, 36 having been constructed by the Air Ministry contractors and 14 by US Army engineers.

Fog Dispersal

FIDO (Fog Investigation and Dispersal Operation) was a system used at certain airfields for dispersing fog and dense smog so that aircraft could land safely. Developed by Arthur Hartley (see Chapter 6) in conjunction with the University of Birmingham, FIDO comprised pipelines on both sides of a runway into which petrol was pumped and then released through burner jets positioned at intervals along the pipelines. These were then ignited by a Jeep with a flaming brand lashed to its rear, and the heat from the flames evaporated suspended fog droplets and produced walls of fire that could be seen from a great distance. This enabled aircraft to have suitable visibility for finding the airfield and landing. Hartley and four others were jointly awarded £5,000 by the Royal Commission on Awards to Inventors (awards 125–127).

Aircraft Production

One of Churchill's 1940 creations was the Ministry of Aircraft Production led by the dynamic Canadian, Lord Beaverbrook. Despite the German attacks on British industrial cities, aircraft production in Britain actually increased from some 15,000 planes in 1940 to 23,000 in 1942. Much of the additional production took place in the shadow factories that had been erected in the North West prior to the war and which were beyond the normal operational range of many of the *Luftwaffe* bombers.

RAF Aircraft

By July 1944 the RAF was at its largest with 487 squadrons (100 of which were provided by the Commonwealth), 7,300 aircraft, and nearly 1,200,000 personnel, of whom approximately 10% were in the WAAF. The ground crews embraced 190 different crafts and trades.

During the Second World War it flew 24 different types of bomber aircraft ranging from the twin-engine Vickers Wellington and Bristol Blenheim that were operational at the outbreak of war to the four-engine Handley Page Halifax and Avro Lancaster.

First used by the RAF in 1942 the Avro Lancaster was a mid-wing cantilever monoplane with an oval all-metal fuselage. It became Bomber Command's principal heavy bomber and flew in over 150,000 sorties. Designed by Roy Chadwick, (see Chapter 3) it was powered by four Rolls-Royce Merlin, or sometimes Bristol Hercules, engines. A 10 metre unobstructed bomb bay meant it could carry the largest bombs used by the RAF, including the 10,000 kg Grand Slam bombs. The crew included the captain, navigator, flight engineer (who sat alongside the captain), bomb aimer, and air gunners.

Flight engineer Sergeant Norman Jackson RAF, a veteran of 102 missions, was awarded the Victoria Cross in July 1944 for going on to the wing of a Lancaster to fight an engine fire while returning from a bombing raid. Group Captain Leonard Cheshire RAF, who had also completed many bombing missions, also received his VC from King George VI on the same day. Protocol meant that Cheshire, as the senior recipient, should receive his award first but Cheshire insisted that despite the difference in rank they should approach the King together. Jackson remembers that Cheshire said to the King, 'This chap stuck his neck out more than I did – he should get his VC first!'

Mosquito Aircraft

Among the senior personnel at the Air Ministry responsible for deciding how resources should be allocated was Air Marshal Sir Wilfrid Freeman, who had played a major part in the development of many aircraft such as the Hurricane, Spitfire, Wellington, and Lancaster. It was he who, despite much opposition, advocated the development of the plywood-framed, twin-engine de Haviland Mosquito. Known initially as 'Freemans Folly', it carried no cannon or machine guns, had a range of 5,600 km, a ceiling of 11,000 metres, a cruising speed of 500 km/h, and a top speed of 680 k/h. It became the most versatile plane of the Second World War and was used in a variety of roles including as a pathfinder aircraft, bomber, night fighter, submarine killer, and high-level reconnaissance aircraft.

It had a high altitude role in the development of the OBOE bombing aid, was part of the precision roof-level attack on the Gestapo headquarters in Oslo in September 1942, and bombed targets in support of the Normandy operations.

USAAF Aircraft (and British Engines)

The Boeing B-17 'Flying Fortress' was the principal aircraft used by the USAAF in its strategic daylight bombing campaign against military and industrial targets in Germany and the occupied territories. It was a four-engine heavy bomber aircraft developed in the 1930s for the United States Army Air Corps (USAAC); despite some early problems, it evolved into the USAAF's main weapon in the European war.

Also used in large numbers by the Americans was the four-engine Consolidated B-24 Liberator aircraft which, although preferred by commanders to the B-17, was not liked as much by air crews. It was also used by RAF Coastal Command.

The USAAF bombers primarily operated from bases in eastern England but a few also flew from bases in northern Italy. In the early days aircraft losses were high, partly because the US commanders and air crews were in action for the first time and also because the *Luftwaffe* had improved its tactics and was operating near its bases. Additionally, many of the American targets were beyond the reach of the short- and medium-range fighter-escorts that could protect them nearer Britain.

The Americans were also working on the Lockheed P-38 twin-engine Lightning (nicknamed 'fork-tailed devil' by the *Luftwaffe*) and the Republic P-47 Thunderbolt as fighter escorts.

Among the American-built aircraft then being received by the RAF was the P-51 Pursuit-Fighter which had been designed and built by North American Aviation Company in response to a specification issued by the British Air Ministry. Test flights indicated that it was not a great performer; however, one of Rolls-Royce's senior test pilots, Ronnie Harker (see Chapter 4), believed it was underpowered.

Fortunately, he was able to substitute a Rolls-Royce Merlin engine for test flying. Witold Challier, a Polish mathematician and engineer at Rolls-Royce, calculated that it would now be able to outperform the latest version of the Spitfire – and he was correct. British engineers at Rolls-Royce and the RAF, together with some UK-based American airmen endeavoured to convince the American designers that substituting Merlin engines would dramatically improve the performance of the P-51.

The Americans took some convincing, though it was eventually agreed that Merlin engines made by Packard should be fitted to USAAF as well as RAF P-51 planes. An unexpected bonus of the substitution was that the planes were now much more economical in fuel consumption, thereby increasing their range. Renamed Mustangs, they became the principal long-range fighter escorts for the US bombers.

Another new development at that time was the introduction of drop tanks which also dramatically increased an aircraft's range. The British introduced a variant made out of stiffened paper rather than aluminium. Drop tanks were also fitted on other fighters protecting the US aerial armadas (as well as on other British and American aircraft). Now protected by long-range Spitfires, Mustangs, Lightnings, and Thunderbolts, it was possible to provide protection deep into Germany and losses to US bombers therefore fell dramatically.

Bombing Aids

Although the British bombing offensive was seen by the public and, hopefully, Stalin as carrying the war to Germany, researches by statisticians reporting to Lindemann established that only 5% of RAF bombers got within 8 km of their targets. Aerial reconnaissance confirmed that the British bombing offensive was lamentable in the early years of the war.

The research into technological aids to improve bombing accuracy was stepped up and two separate electronic systems were developed. The first, which had two variants called Gee and Oboe, was based on the principle of triangulation and involved sending radio signals from two transmitters in the UK. Equipment aboard the plane could measure the time interval between the receipt of the two signals, thereby enabling the plane's position to be calculated. Gee had a maximum range of 500 km and thus was ineffective over the many German cities that were beyond that distance from the transmitters, but Oboe could operate at a longer range. Both systems were, however, vulnerable to jamming by German transmitters based in Holland.

The alternative H2S system had small radar sets installed in the aircraft which could map the terrain, and hence the targets, up to 50 km ahead of the plane. ASV (air to surface vessel) equipment was already being used in the search for U-boats and Bomber Command adopted a version of this as well as a new H2S system. The operation of both these systems came from the development of the cavity magnetron and centimetric radar. H2S was developed by a team at the Telecommunications Research Establishment whose leader, A. D. Blumlein, was tragically killed when the plane in which he was flying to test new equipment crashed in Herefordshire in June 1942.

Blumlein, Alan (1903–42)

An electronics engineer, he was considered one of the most significant engineers and inventors of his time, receiving 128 patents in the fields of telecommunications, sound recording, stereophonic sound, television, and radar. In the Second World War he was seconded from Electrical & Musical Instruments (EMI) to the Telecommunications Research Establishment and led the team which developed H2S, a navigational and bombing aid widely used by the RAF and other air forces and which continued in use well into the later years of the 20th century. He was killed when the Halifax aircraft in which he was flying to undertake research into the further development of H2S crashed in Herefordshire in June 1942.

The Dam Busters Raid (16–17 May 1943)

In addition to carpet bombing German targets, Bomber Command decided to bomb the Eder, Mohne, and Sorpe dams in the Ruhr Valley. This was despite advice from the Ministry of Economic Warfare that damage to the dams chosen would not seriously affect the production of war material.

Dr Barnes Wallis was chosen to investigate the practicalities of achieving this and he devised the now-legendary cylindrical 'bouncing bomb' which was given a rotary motion before leaving the plane. When the bomb hit the surface of the water it replicated the bouncing of a spinning stone thrown onto a flat surface of water. Extensive trials were first carried out at Chesil Beach, Dorset, prior to confirmation by further tests at British reservoirs.

No. 617 Squadron was an elite squadron created to undertake the task. Their Lancaster bombers had VHF radio telephones installed and their bomb bays modified to carry and apply a spin to the bombs.

Wing Commander Guy Gibson RAF was chosen to lead the raid, and he personally selected the crews. They all underwent intensive training, first in very low level night-flying over water, then bomb dropping from exact heights at exact speeds at the Derwent Reservoir in the Peak District and the Elan Valley reservoirs in mid-Wales. In one of the dress rehearsals six out of 12 aircraft were seriously damaged. Ground crews worked extremely hard to provide the necessary support.

The Sorpe Dam was only slightly damaged; it is now known that despite Bomber Command's high expectations the resulting damage to German war production was less than anticipated. Sadly, eight of the 19 aircraft that participated failed to return.

Wing Commander Gibson was awarded the VC and 33 other members of the Squadron were also decorated.

Tallboy and Grand Slam Bombs etc.

For daylight precision bombing raids against particular targets Wallis designed a variety of heavy bombs including the 12,000 lb (5,400 kg) Tallboy and the 22,000 lb (10,000 kg) Grand Slam earthquake bombs which were used against V1 launch sites, U-boat pens and other highly rated targets in France, and the *Tirpitz* battleship in Norway. The Americans also developed their own vertical bombs including the AZON and RAZON which used the azimuth and range guiding systems to improve targeting accuracy.

Countering V1 Flying Bombs and V2 Rockets

Covert eavesdropping on captured German generals had informed the Allies that Hitler had a secret weapon, *Wunderwaffen*, which it was believed could, even at that late stage of the war, transform German fortunes. In addition, intelligence reports from agents and aerial reconnaissance had located sites in northern France and Germany where unusual activity was occurring. Aerial bombardment by the Allies did not destroy these sites but did delay the introduction of the new weapon.

Exactly one week after D-Day the Germans launched their first V1 flying bomb from the Pas-de-Calais; this was followed by nearly 10,000 more aimed at London and the South East over subsequent weeks. Also known as 'buzz bombs' and 'doodlebugs', at their peak they were being launched at a rate of 100 a day. The RAF and USAAF bombers attacked their launch sites which had to be moved eastwards as the Allied armies advanced across France. The last bombs aimed at Britain were launched in October. After this a further 2,400, fired from Germany, were aimed at the port of Antwerp and other targets in Belgium, the attacks only stopping when the last launch site was overrun at the end of March 1945.

The V1 (the world's first unmanned aerial plane) comprised a pilotless plane containing high explosives; guided by autopilot, it was powered by a single pulse jet engine. The plane's direction of flight was determined when it was launched. After flying a predetermined distance (or when it had run out of fuel) it dived to the ground and exploded. When people on the ground could hear the noise of the engine they knew they were safe, but as soon as the noise stopped they dropped everything and made for the nearest shelter – frequently the cellar of a house they happened to be passing.

Whilst flying the planes were attacked by anti-aircraft guns and RAF fighter planes that sought to shoot them down or, in a few cases, overturn them by touching and raising the undersides of their wings! In addition to attacking launch sites, Bomber Command also attacked the underground storage depots and the factory in which they were made in the small village of Peenemünde on the Baltic coast.

Altogether over 9,500 flying bombs were aimed at London and more than 2,500 at targets in Belgium. Casualties were mostly among the civilian population and over 5,500 persons were killed in South East England. Coming as they did at the end of five long hard years of war and when an Allied victory was believed to be within sight, the V1 attacks caused a significant drop in British morale. But their effect was as nothing compared to the consternation caused by the arrival of the supersonic V2 rockets, which killed thousands of people and further shattered the euphoria of the British population (see Appendix 8).

For launching the rockets the German military favoured a number of mobile launchers but Hitler insisted on a single major launch complex at La Coupole, 5 km from Saint-Omer, Pas-de-Calais, which was constructed with a domed concrete roof 71 metres in diameter and 5 metres thick. Attacked by the RAF shortly after completion, the dome withstood the onslaught though the access tunnels were destroyed; this rendered the complex unusable, thus making the use of mobile launchers the only possible option.

It has been suggested that German spies who had been captured and 'turned' were ordered by British intelligence to send messages to Germany saying that the initial bombs were landing northwest of London in the hope that the Germans would then reduce the range settings of future rockets, thereby causing them to land southeast of London. Although this has never been officially confirmed there is strong anecdotal evidence that a successful disinformation campaign was in fact mounted.

The British authorities first learned that rockets were being manufactured at Peenemünde when a German scientist, Hans Mayer, who was concerned at the direction Germany was heading, posted what became known as the Oslo Report to London when he was staying at a hotel in Oslo in November 1940. While most senior people in London thought the report was a plant, Dr Jones, (see Chapter 4) the senior scientist at the Air Ministry, thought otherwise.

Then, by 1943, British intelligence had received various secret reports confirming that production was actually taking place there. After aerial photo reconnaissance it was decided to launch a bombing raid on the site. This took place with Lancaster and Halifax bombers on a moonlit night in August 1943. Although the site was not seriously damaged, the Germans now knew its presence was known to the British and production was moved deep inside Poland where it would be harder to bomb. Some 12,000 forced labourers and concentration camp prisoners were killed producing the weapons. Despite the best efforts of the British military and of top scientists and engineers no effective counter to the rockets was discovered – nevertheless the disinformation campaign did reduce their impact. Also, using data about the rockets' launch curves obtained from radar stations, Allied plotters with slide rules could rapidly locate a launch site's position which was then immediately attacked by Mosquito bombers.

Jet-powered Fighters

The Gloster Meteor was the first British jet fighter and the Allies' only operational jet aircraft during the Second World War. Its development was based on the innovative turbojet engines conceived by Frank Whittle and Power Jets Ltd. (see Chapter 4). Development of the aircraft itself began in 1940, though work on the engines had been under way since 1936. The Meteor first flew in 1943 and commenced operations on 27 July 1944 with No. 616 Squadron RAF. Nicknamed the 'Meatbox', it was a successful combat fighter, despite not being a sophisticated aircraft in its aerodynamics, but entering service late in the war meant it saw only limited action.

Appendix 6

Mobilisation of British Industry

Nine new shadow factories – aircraft factories built alongside, and staffed by those from, their motor vehicle counterparts – had been constructed in the Midlands and some northern counties. This was part of the pre-war rearmament programme to increase the manufacturing capacity of the aircraft and aero-engine industries in the event of war. They were developed by the Air Ministry but not equipped or staffed until war started, when they then became the operational responsibility of the British aircraft and motor car industries. Herbert Austin, the founder of the eponymous motor car, was in overall control. In addition, extensions were built to the factories of the principal aircraft and aero-engine manufacturers so as to facilitate expansion should the need arise.

British engineers, together with their colleagues in other disciplines, designed and built the new factories and military establishments; researched, designed, and developed new products; switched operations from civilian to military output; increased production rates; and manufactured munitions, aircraft, ships and fighting vehicles in ever-increasing quantities. Other engineers designed the Anderson and Morrison domestic air-raid shelters for installation in gardens and indoors respectively.

Generally, most British ships, aircraft, artillery, tanks and other vehicles were developed and built by the country's major industrial manufacturers, who were often working in association with the Government's principal scientific research and development departments. Others were developed in the Royal Arsenal (in Woolwich) or in the Royal Ordnance Factories. Armoured and other mechanised vehicles were designed and made by the specialist departments of the traditional manufacturers of civilian vehicles to specifications set by the War Office.

Although most of the shipyards and many of the factories needed for the manufacture of the weapons, equipment, and machines required to mount the increasingly intensive war effort had been built pre-war, the construction of many more began following the start of the war, with some 300,000 building workers being employed.

Munitions

For over three centuries cannons, shells and explosives for both the Army and the Navy had been manufactured at the Royal Arsenal and the Royal Gunpowder Mills, Waltham

Abbey; since the 1800s, rifles and other small arms had been made at the Royal Small Arms Factory, Enfield. These were expanded and continued to operate during the war and, in addition, a number of Royal Ordnance Factories (ROFs) were established northwest of a line from Somerset to Northumberland. They had to be sited where there were good transport links and a plentiful and guaranteed supply of clean process water. For safety reasons they also had to be located away from centres of population but where there was an adequate work force within reasonable travelling distance.

Many of the new ROFs were designed and constructed by the Royal Arsenal while others were built by the Ministry of Supply, the Ministry of Works, and two private companies. ICI also built and owned a number of munitions factories and some were operated by private companies previously unconnected with the armaments industry, for example Imperial Tobacco Co. Ltd, Courtaulds Ltd, the Co-operative Wholesale Society, Metal Closures Ltd and Lever Brothers.

The Bren gun, Britain's primary light machine gun in the Second World War, was so named because a Czechoslovakian-designed gun manufactured at **BR**no was modified by engineers at **EN**field. The Sten, a simple submachine gun (four million were made during the 1940s), was designed by engineers Reginald **S**hepherd and Harold **T**urpin from the **EN**field factory.

After the war started the output of factories that produced engineering products was switched to those required for the war. Production rates actually increased for most engineering factories, large and small, despite the loss of some key staff to the armed forces. Other factories which in peacetime were involved with non-engineering products switched to more technical articles as well.

Ford Motor Company

In addition to Dagenham, production was undertaken in new factories in Manchester and Leamington. With some 3,000,000 ha being added to Britain's arable land, many more tractors were needed to produce the home-grown food required to reduce food imports. Ford manufactured some 120,000 tractors, which represented 85% of all tractors working in Britain by the end of the war. Over 250,000 V-8 engines were manufactured and used to power every light-gun carrier and many motor torpedo boats and landing craft, for hoisting barrage balloons, and as standby generators at RAF stations. Over 30,000 Merlin aero-engines were manufactured at the company's Manchester factory. The company also developed improved means of waterproofing vehicles prior to the Normandy invasion and constructed General Montgomery's trailer caravan which he used as his operations room. It is a tribute to the company's workforce that output rose during the *Blitz* in spite of the effects of bombing on industrial, civilian, and infrastructure assets. When the threat of invasion in 1940 was high, a secret scheme for the destruction of major assets was drawn up.

Marconi Electrical Company

With headquarters at Chelmsford the Marconi company manufactured a whole range of small electronic equipment including: magnetrons for centimetric radar, of which several thousand were made each month; naval radar sets; small suitcase-mounted radio transmitters for use in lifeboats; submerged buoys dropped from spotting aircraft to inform RN submarines of the location of enemy minefields; and mobile radio units which were operating within 12 hours of an invasion landing. Of the equipment sent to Southeast Asia some 80% was lost either by sinking when in transit or by damage due to the tropical weather. Experience demonstrated that packaging was of paramount importance, equipment had to be stored and transported upright, and fungus could be controlled by varnishes. Radio sets used by the Chindits had to be waterproofed by being encased in metal. External projects comprised duplication in Dorset of the country's main Daventry radio transmitters; the transmitters for a 'phantom' GHQ Liaison Regiment; control equipment for the D-Day landings installed in a bunker 30 metres below ground; and 16 transmitters near Deal, Kent, to jam the radio signals thought to control the V2 rockets. Nearly 10,000 Marconi operators, of whom some 1,000 were lost, served as radio officers in merchant ships. In January 1945 Marconi fitted out SS *Franconia* with high-level communication systems for the Yalta conference.

Metropolitan Vickers

From the main factory at Trafford Park, Manchester, and associated sites, the company was responsible for constructing much of the electrical and mechanical equipment, as well as the Lancaster bomber. They made other, more routine products, such as engines, generators, degaussing equipment and geared turbines for ships, searchlights and sound locators, radar equipment for searchlights and gun laying, automatic pilots for aircraft, a special non-magnetic steam-driven drill for piercing the casing of unexploded bombs, and buoys for marking the positions of U-boats. The research department was also involved with high-voltage developments, penicillin production, X-ray spectography, the cyclotron, and various metallurgical investigations.

Vauxhall Motors Limited

From the company's main factory in Luton a wide range of vehicles and other equipment was manufactured, including Churchill tanks, gun mountings and welding sets. They also carried out special projects – assistance with jet engines; waterproofing of invasion craft; and 'giraffes', a reverse of the double-decker bus principle whereby the driver and engine were on top for possible use with the invasion fleet if waterproofing of conventional vehicles should prove unsuccessful – to name but a few. Following the invasion at Pearl Harbor rubber was not available and wheels, based on those of dodgem cars, were made from cotton duck.

Butterley Company

Typical of medium-size companies was the work and output of the Butterley Company of Ripley, Derbyshire. With their main activities being the operation of collieries and brickworks, and the manufacture of miscellaneous engineering products, they were well placed to diversify into military and naval projects. They manufactured, amongst other things, keels for 51 frigates, pontoon units and floats for the Mulberry harbours, Bailey bridge panels, and cannon and tank turrets. They also developed an alternative portable bridge system for use in case the Bailey bridge was not successful. During the war the factory area was camouflaged, women were employed for the first time, and a medical service was established.

Appendix 7

POWs' Ingenuity

The ingenuity shown by Allied prisoners of war (POWs) in their attempts to escape made an indirect contribution towards victory, not only by increasing the numbers of enemy troops assigned to guard them, but also by raising the morale, both of the prisoners themselves and of the population at home. The technical improvisations adopted by POWs in their attempts to escape were often extraordinary

POW Tunnels

Most escape attempts were made by digging tunnels under the perimeter fencing, and many ruses were adopted to conceal the prisoners' activities from their guards. If the

Colditz Castle

Colditz Castle is perched on a cliff overlooking the town of Colditz in Saxony, Germany. It was used as a high-security POW camp known as Oflag IV-C to which many recaptured POWs were sent. Despite being believed by the Germans to be 'escape proof' there were over 35 successful escapes by British, Commonwealth, and Allied officers. The first of the 16 successful British escapees was the former sapper Lieutenant Airey Neave RA (1879–1961) who got back to the UK in April 1942. Thereafter he resumed service with the Army and was awarded the MC for his escape. By the end of the war he had also been awarded the DSO and been appointed MBE. He later became an MP and, when a member of Margaret Thatcher's Shadow Cabinet, was assassinated by Irish Republicans in the House of Commons car park in 1979.

Another successful British escapee was the civil engineer Lieutenant Pat Reid RASC (1910–90). After two unsuccessful attempts to escape from Colditz he became the British POWs' 'Escape Officer', responsible for overseeing all British escape plans. As such he assisted in many escape attempts until April 1942 after which he was able to organise his own escape – he managed this in October 1942 together with three fellow officers. On reaching Switzerland he was recruited by MI6 and stayed there gathering intelligence from other returning POWs. He had been awarded the MC for bravery during the Battle of France in 1940 and was appointed MBE in 1945. Other escapees reaching Switzerland also had to stay there because there was no means of returning to the UK; some spent time skiing and in other leisure pursuits.

The Three Colditz Musketeers

Major Anthony (Tony) Rolt MC & Bar (1918–2008) was, both before and after the war, a leading British racing driver and support engineer. Having started his racing career in 1935 as a 16 year old driving a Morgan 3 Wheeler in a speed trial, he won the British Empire Trophy in 1939 and the Le Mans 24 Hours race in 1953. He also participated in three Formula One World Championship Grands Prix races. After retiring from competitive driving he concentrated on developing transmission systems for four-wheel-drive vehicles, and his company became a major technology partner of Ford, Chrysler, Audi, Fiat, and General Motors. He had been awarded the MC for gallantry during the defence of Calais in 1940. After being taken prisoner he made several determined escape attempts from the various camps to which he was taken, and on one occasion was within metres of the Swiss border when he was recaptured. After the war he was awarded a bar to his MC for his various escape attempts.

Flight Lieutenant Bill Goldfinch RAF (1916–2007), a former 2nd lieutenant in the Royal Engineers, was the pilot of a Short Sunderland flying boat that was holed and sunk after hitting an underwater object when landing in the dark during the evacuation from Greece. Badly injured, he was taken to a hospital which was later overrun by the Germans.

Flight Lieutenant John (Jack) Best MBE RAF (1912–2000), who had been captured after his plane had run out of fuel, also off Greece, met Goldfinch in the Greek hospital. After they had recovered from their injuries they were initially interned at the POW camp Stalag Luft III. They and a fellow prisoner escaped, though they were recaptured after reaching Poland and were then sent to Colditz. Best attempted to escape several times but was recaptured on each occasion. From April 1943 to March 1944 he became a 'ghost prisoner', hiding under floorboards and in closets in Colditz to trick German guards into believing that he had escaped.

tunnels were excavated in unstable ground, basic engineering knowledge was required to construct the structural support needed to prevent its collapse.

The Colditz Cock

Probably the most audacious of all the wartime attempts by prisoners to escape from captivity was the Colditz Cock, which involved the construction of a glider in one of Colditz Castle's attics. Although a number of other prisoners played significant supporting roles, the three POWs principally involved were the Three Musketeers: Major Tony Rolt, Flight Lieutenant Goldfinch, and Flight Lieutenant Jack Best.

Rolt's idea of building a glider in which Goldfinch and Best would make an airborne escape from the castle was conceived following the discovery in the prison library of *Aircraft Design*, a two-volume work that explained the necessary physics and engineering and included a detailed diagram of a wing section. Geoffrey (later Air Commodore Geoffrey) Stephenson assisted in the project.

Made of wood and cloth acquired by various subterfuges, the glider was assembled by Goldfinch, Best, and 12 assistants known as 'apostles' in a screened-off part of an attic above the castle chapel in one of the towers. A 20-metre-long runway was to be constructed from tables assembled on part of the roof invisible to the guards. Weighing 109 kg, the glider measured 6 metres in length and had a 9.8 metre wingspan. It was to be launched using a catapult that consisted of a bathtub filled with concrete at the end of a series of cables and pulleys. As the bathtub fell to the ground, the cables would fling the glider into the air, where it would float for about a mile over the town of Colditz and the Mulde River.

From there Best and Goldfinch were planning to walk to a train station and escape by train to Switzerland. Before they could launch the glider, the camp was liberated by the US Army on 15 April 1945. In 1999, Best was a consultant for the Channel 4 documentary *Escape From Colditz* in which a full-scale replica of the glider was built and successfully flown. The book, *Flight from Colditz*, was published in April 2016.

MI9, MIS-X, and BAAG

Military Intelligence Section 9 (MI9) was a War Office department tasked with facilitating the escape of British POWs and the return of those who had escaped. MI9 used the services of magician Jasper Maskelyne to design hiding places for escape aids, including tools hidden in cricket bats and baseball bats, maps concealed in playing cards and actual money in board-games. An escaper's knife was designed which incorporated a strong blade, a screwdriver, three saws, a lock pick, a forcing tool, and a wire cutter.

Among the escape aids which RAF aircrew had incorporated into their uniforms or which were sent to POW camps were: silk handkerchiefs on which maps has been printed; boots with detachable leggings that could quickly be converted to look like civilian shoes; and hollow heels that contained packets of dried food. Compasses were hidden inside pens and tunic buttons. Left-hand threads were used so that if the Germans discovered them and tried to screw them open, they would just tighten. Some of the spare uniforms sent to prisoners could be easily converted into civilian suits. POWs inside Colditz Castle requested and received a complete floor plan of the castle. Forged German identity cards, ration coupons, and travel warrants were also smuggled into POW camps by MI9.

MI9 sent the tools in parcels in the name of various, usually non-existent, charity organisations. They took care not to use Red Cross parcels lest they violate the Geneva Convention and to avoid the guards restricting access to them. MI9 relied either on their parcels not being opened or the prisoners (warned by a message) removing the contraband before they could be searched. In time the guards learned to expect and find the escape aids.

The American equivalent organisation was MIS-X which was based in Fort Hunt, Virginia. It sent similar devices to American POWs. A radio code was devised to send messages via BBC broadcasts that were based on the Morse code preceded by a bell tone. A method was developed of hiding a message in innocuous-looking outgoing letters to the USA. To avoid compromising legitimate aid organisations like the Red Cross, MIS-X invented its own aid organisations, creating aid packages in which secret items were hidden.

In Southeast Asia the British Army Aid Group (BAAG) based in southern China helped British and Allied POWs escape from Japanese captivity.

Appendix 8

Combating Enemy Technology Brains Outmanoeuvring Brains

Germany, and to a lesser extent, Italy and Japan, developed various new weapons, the V1 flying bombs and V2 rockets being notable examples. Many Allied scientists and engineers were engaged in developing countermeasures, often with limited information. In addition, many resistance fighters and Allied agents endeavoured to discover what was being undertaken in various clandestine operations.

As the war dragged on it became a grim battle for survival for Germany. One or two *Wunderwaffen* were indeed produced, but always on a highly selective basis and depending on the *Fuehrer*'s whim. Anything that did not fit with his view of how land battles should be fought got short shrift.

In the interwar years the German High Command devoted a lot of effort to analysing the lessons from the First World War, leading to two seminal books, *Achtung Panzer!*, by Heinz Guderian, and *Infanterie greift an*, (The Infantry are Attacking) by Erwin Rommel. (Both are still available in English translation.) In the first, Guderian examined the use of tanks by the British at Cambrai in 1917 where, according to him, the British convincingly demonstrated that the arrival of the tank meant the end of trench warfare. He argued that the British failed to press home their advantage at Cambrai and had not drawn the right conclusions about what the advent of the tank meant for the future of land warfare. His book was an attempt to do so, and it established the intellectual foundations of the *Blitzkrieg*. Given its central location, Germany was in no position to withstand a long war, and the *Blitzkrieg* idea of swift, one-off attacks fitted well with this. It was, however, a less vital concept for peripheral powers such as Russia and Britain; the deeper resources at their disposal would give them the upper hand in a long war.

The German approach to the use of technology in supporting their war effort was in complete contrast to that of Britain and the USA. In 1940, after the fall of France, the Germans believed that they had already won the war. As a result scientists were conscripted into the armed forces and it was ordered that no scientific research or development should be pursued unless it would be of use within four months. These policies were not reversed until late 1942, by which time the Allies had an immense technological advantage as well as greater reserves of assets and manpower.

German scientists and engineers displayed their traditional ingenuity despite the relatively poor relations that existed between the German military and the country's scientists; the difficulties they encountered from Hitler's frequent stop-start instructions; the shortages of fuel and key materials that resulted from the Allied blockade; and the Allied air offensive. Against these odds, they invented and manufactured some remarkable machines and weapons which involved a range of innovations that spanned a wide spectrum of technology.

In addition to the V1 flying bombs and V2 rockets, these included the V3 supergun; radar; the German jet powered plane (which actually flew before the RAF's Gloster Meteor); inventions to increase the effectiveness of the U-boat fleet; the Atlantic Wall; and various synthetic materials to replace those no longer available or in short supply because of the Allied blockade.

While Hitler hoped that these 'secret weapons' would turn the tide of the war, all were developed too late to make a difference to the eventual outcome. The V1 flying bombs and V2 rockets which in the summer of 1944 caused such consternation and disbelief in Britain – temporarily shattering the euphoria of a nation that, despite its war weariness, believed it was within sight of ultimate victory – only demonstrated what effect they might have had if they had been deployed earlier.

Throughout the war institutional hubris dogged the Nazi leadership, which was repeatedly wrong-footed both by Allied technological initiatives and by the reluctance to believe that if they had not built a particular weapon or device themselves their enemies must lack the wit to have done so. (Lord Cherwell was guilty of the same failure when V2 rockets were landing on London – he could not believe that the Germans could have developed such a thing.)

Sources which have given detailed descriptions of the many enemy weapons include Roger Ford's *Germany's Secret Weapons of World War II;* the two books entitled *Secret Weapons of World War II* by William Breuer and William Yenne*;* Brian Ford's *Secret Weapons;* National Geographic's TV series *Nazi Megastructures*; and various pages online.

Paratroops and Blitzkreig *Targets*

In their 1940 campaigns against the Low Countries and France, the Germans pioneered the use of both paratroops and *Blitzkrieg* tactics. All of these required specially designed equipment and, the latter in particular, the use of sophisticated and rapid communication systems.

The losses sustained by the German paratroops invading Crete were so great that Hitler banned their use thereafter, but the British and American armies adopted them and they were successfully deployed in the D-Day landings. They had also been used, albeit less effectively, in the Sicilian and Salerno landings and at Arnhem. The *Blitzkrieg* concept was also adopted by the Allied armies in north-west Europe.

V1 Flying Bombs

A proposal for a pilotless plane had been submitted by the aero-engine company Argus before the war but rejected by the German Air Ministry for various reasons. By the spring of 1942 the Germans realised they were not winning the air war and the idea was revived and given top priority. The flying bomb was to be launched from an inclined ramp in northern France and to be powered by a propulsive charge of hydrogen peroxide and permanganate. It incorporated an automatic pilot and an electro-mechanical device to cut out the engine when it had covered a predetermined distance, thus bringing it and its explosive warhead down. Testing and development were carried out at the rocket research and testing station at Peenemünde on the Baltic coast. Their effectiveness and the countermeasures adopted are described in Chapter 9 and Appendix 5.

V2 Rockets

As early as 1932, when he was only 20, Wernher von Braun (1912–77) started experimenting with rockets. Shortly after Hitler came to power the German military began funding his work. Told in 1935 to find a suitable deserted site, he chose Peenemünde at the apex of a bay on the Baltic coast, where a 25 km^2 secret base was established complete with its own power station and a supersonic wind tunnel capable of generating a Mach 4 wind speed. After the fall of France, Hitler agreed that work should continue but only with a low priority as it was assumed that conventional weapons were all that would be needed to win the war.

Experimental launches continued over the next three years; however, multiple problems occurred, and although each flight was photographed from locations around the coast, existing technology made it difficult to establish the cause of the malfunctions. Two of the major problems were: designing a liquid oxygen-powered engine with sufficient boost to make a 13-ton missile reach the Mach 4 speed it needed to travel 200 km, and devising the technology that could keep it stable and on course. The British, having become aware of what was going on at Peenemünde, carried out an attack by the RAF in May 1943, and while the physical damage was not great, some of the leading technologists were killed.

In July 1943 Hitler ordered that almost impossible numbers be built, and a new underground manufacturing site was hastily constructed by slave labour in an old mine at Mittlewerk in Central Germany. Comprising a 3 km-long tunnel in which the manufacturing was undertaken, together with a parallel one for logistics and numerous other interconnecting passageways, the complex totalled 11 km of tunnels in all and was capable of producing 900 rockets a month. Some 10,000 prisoners worked and, in many cases, died there. Mobile launchers had to be used following the RAF attack on the complex at La Coupole, and the rockets – each measuring some 14 metres long by 1.65 metres wide – were transported to the sites by rail.

The first successful launch on London was on 8 September 1944, and over the following six months some 3,000 were aimed at the UK and, later, the Low Countries, killing around 5,000 people. Travelling at over 5,500 km/h – more than twice the speed of sound – and dropping to the ground from a height of around 90 km one would land at its destination before its sound waves. Therefore, the first that people in the vicinity knew that their neighbours had been attacked was when they heard the noise of the rocket exploding. Had the Germans been able to launch them six months earlier, the last year of the war could have been different and the Normandy invasion might well have been affected.

Although the British and Americans had agreed that post-war they would share the secrets of the V2, von Braun was captured by the Americans shortly before VE Day. Together with about a hundred scientists and their families (and a hundred rockets), he was taken to the US. There he later led the American space programme that included the first moon landing in 1969 by the American Apollo 11 rocket.

V3 Super Gun

Another of Hitler's so-called secret weapons was the V3 super gun, located in a heavily reinforced bunker deep underground at Mimoyecques near Boulogne. It was aimed at the centre of London in the hope that it would add to the chaos being created by the V1 flying bombs and V2 rockets. To achieve the velocity necessary to reach London, German engineers devised an ingenious 25-barrelled long cannon which incorporated a series of propellants along the length of the barrel. Fortunately it did not become operational in time and was never fired, although a similar but smaller cannon fired shells from the Pas-de-Calais to Dover regularly from 1940 to 1944, causing relatively little damage.

Because the site was located deep below ground in a heavily reinforced bunker, it was difficult to destroy by conventional means. The Americans devised one of the world's first ever drones (a remote-controlled heavy bomber packed with explosives) to attack it, while the British used Barnes Wallis' earthquake-creating bomb. Such bombs contained large amounts of explosives and were dropped from a high altitude, enabling them to penetrate deep underground before exploding and causing shockwaves. After being first disabled by Allied bombing, the V3 was finally destroyed by sappers of No. 5 Bomb Disposal Company in November 1944. Dr Hugh Hunt of the University of Cambridge is currently undertaking model experiments to discover more about its design and operation.

Radar

The Germans had developed their early warning *Freya* radar system (named after the Norse goddess *Freyja*) before the British, but even though their radar systems were technically more advanced they did not link their stations to an effective ground-to-air detection and control system – and never imagined that Fighter Command might have done so. The Germans built a thousand radar stations, and a naval version operating on a slightly different wavelength was developed as the *Seetakt*.

In February 1942 aerial reconnaissance detected one of their stations at Bruneval, Normandy, and shortly afterwards the site was attacked by British airborne commandos. The key elements of the German equipment were dismantled and, along with a German radar technician, taken to the UK by boat, thereby enabling British scientists to understand the technology behind the German equipment. Information later supplied by the French resistance confirmed that all German sets operated on a small number of frequencies which could be easily jammed.

For the bombing attack on Hamburg in July 1943, the RAF rendered German radar completely ineffective by dropping metal strips known as 'window' or 'chaff' (see Chapter 4). Although the Germans were aware of this possibility, and had their own similar system known as *Duppel*, they had done nothing to counter its use by the Allies. 'Window' was there after used frequently by the Allied air forces, and it played a part in deceiving the Germans on D-Day.

The Atlantic Wall

Stretching from the Franco-Spanish border to Norway and taking in the Channel Islands, the Atlantic Wall was a massive fortification comparable to the Great Wall of China. Made from 17 cubic million metres of concrete and 1.2 million tons of steel, it was intended to help achieve Hitler's dream of ensuring that the Third Reich would last for a thousand years. On 6 June 1944, it was breached in a single day.

According to Albert Speer (1905–81), the architect who was the German Minister of Armaments and War Production, Hitler personally designed the fortifications down to the smallest detail, and even insisted that one-twelfth of the steel and concrete should go to the defences in the Channel Islands because of their propaganda value as a British territory under German control. The wall was built under the direction of the Todt Organisation, harnessing for its construction some 300,000 slave labourers, many of whom died through starvation, overwork or disease.

In addition to the massive gun emplacements and numerous machine gun posts, there were thousands of pillboxes and tank traps, and millions of mines and sharpened stakes known as *Rommelspargel* ('Rommel's asparagus'), which were designed to impale landing paratroopers. There were some 15,000 fortified positions for some 300,000 defenders, and in Normandy alone there were 8,000 firing positions and 3,000 artillery pieces. Some of the extant structures are now tourist attractions, and although they are now mostly in decay, it has been suggested that the Wall's remains should be declared a World Heritage Site.

Rocket- and Jet-powered Aircraft

In his 2015 paper 'German Aircraft Design during the Third Reich' Dr A. D. Harvey, having questioned whether the *Luftwaffe* ever did have the best equipment, said that in 1944 the hard-pressed Germans brought into service two aircraft that were far in advance of anything available to the Allies. These were the Messerschmitt Me 163 *Komet* rocket-powered fighter and the Messerschmitt Me 262 jet fighter. He went on to say, however, that these were issued to combat units long before they were ready for going into action.

The concept behind the Me 163 fighter was that it would gain altitude under rocket power and then operate as a high-speed glider while using bursts of rocket power to maintain its speed. When functioning properly it did indeed handle beautifully; on the other hand, it had a tendency to explode when the rocket was reignited or when it touched down. Over 300 were built and some saw action, but they achieved relatively few victories.

The Me 262 jet plane suffered other problems: it had relatively high take-off and landing speeds with poor throttle control when landing. British experts, when studying the plane after the war, were particularly unimpressed by its Jumo 004 engines which required a major overhaul after 25 hours, compared with the 125 hours for the British jet engine. Sir Frank Whittle thought that the difference was mainly due to the quality of the heat-resisting materials used.

The Me 262 had swept-back wings because the weight of the jet engine made it necessary to move the plane's centre of gravity (not due to advanced aerodynamic theory). The Germans did, however, develop the Junkers Ju 287 bomber which had swept-forward wings.

Air-to-Sea Guided Missiles

The FX-1400 radio-controlled armour-piercing bombs were mounted on Dornier Do-17 bombers and released at a height of about 5,000 metres, enabling them to attain a high terminal velocity. They were used in September 1943 against the Italian fleet when it was sailing to surrender to Admiral Cunningham, resulting in the sinking of the battleship *Roma* and damage to the battleship *Italia*.

Ejector Seats

In 1941 the Germans invented the ejector seats. These are now in more or less universal use for military aircraft.

Submarine Technology

As described earlier, technical improvements to the design and operation of U-boats continued throughout the war. The most significant that went into service were the *Schnorchel* air breathing tube and radar detecting sensors. The former allowed the

U-boats to operate underwater for longer periods, while the latter could nullify the effectiveness of the airborne radar equipment installed in Coastal Command aircraft. The construction of a new generation of more powerful U-boats (Types XXI and XXIII) was hampered by Allied bombing, and they were unable to go into service before the war ended.

Radio Controlled Tanks

In the battle that raged following the Anzio landing in April 1943, a small abandoned German tank was found to be radio controlled. REME and RE troops, operating under enemy fire, were able to recover it for detailed examination and discovered it was to be used for demolition purposes. Its single operator would take it as near as possible to his demolition target and would then dismount and control it by radio for the remainder of its mission.

Synthetic Fuel

The two methods of converting coal to a liquid synthetic fuel had both been invented in Germany, one in 1913 and the other in 1923. During the Second World War they became the country's primary sources of high-grade aviation fuel, synthetic oil, synthetic rubber, synthetic methanol, synthetic ammonia, and nitric acid.

Allied Acquisition of Enemy Technology

After the war Britain, the USA, and Russia sent teams of engineers and scientists into Germany to salvage their secrets, prevent the destruction of sensitive equipment, and recruit (or press gang!) leading engineers and scientists. Among the German scientists who moved to Britain were Dietrich Kuechemann and Kurt Mangler, who joined the RAE at Farnborough and made notable contributions to the development of the Concorde. Both the USA and Russia acquired some of Germany's leading rocket experts.

The end of the war revealed that the Germans were ahead in many areas of scientific research, even though this advantage had not yet resulted in practical applications. In the field of boundary layer transition, for example, German scientists Tollmien and Schlichting had developed at Goettingen the now widely accepted theory that turbulence occurs when small amplitude disturbance waves start to become unstable and grow exponentially. They had also demonstrated this experimentally.

Italian Two-man Submarine

As recounted in Chapter 8, the Italians, who had first developed manned midget submarines, succeeded in damaging two British battleships when in harbour in Alexandria. They had also mounted attacks on ships in Gibraltar harbour, often operating surreptitiously from a base across Gibraltar Bay in the Spanish port of

Algeciras. Limpet mines were frequently attached to the hulls of the ships they were attacking; when discovered, these had to be defused and removed. Among Gibraltar-based naval officers involved in countering such Italian underwater activities was Lieutenant Peter Danckwerts GC RNVR (see Chapter 3).

Japanese Inventions

Among the weapons developed by the Japanese were air-to-surface guided missiles, surface-to-air missiles, rocket-propelled aircraft, and manned torpedoes – in the latter two the *kamikaze* (divine wind) pilot was on a suicide mission. A special aircraft, the MXY7 *Okha* (cherry blossom) was designed for the former. Of the total weight of some 2,100 kg, over half was in the explosive warhead. The power plant was a cluster of three short-duration rockets which enabled the aircraft to attain a speed of nearly 1,000 km/h in a steep dive towards the target warship, making it extremely difficult to shoot down.

Bibliography

Awards

Royal Commission on Awards to Inventors, First report, Session 1948–49, Cmd. 7586, Appendix 1, pp. 19–21.

Royal Commission on Awards to Inventors, Second report, Session 1948–49, Cmd. 7832, Appendix 1, p. 10.

Royal Commission on Awards to Inventors, Third report, Session 1952–53, Cmd. 8743, Appendix 1, pp. 11–13.

Royal Commission on Awards to Inventors, Fourth and final report, Session 1955–56, Cmd. 9744, Appendix 1, pp. 16–18.

General Histories, Miscellaneous etc.

Bourne N. (2010), 'Rhydymwyn Valley Works: Lifting the lid on secret site' BBC. See http://news.bbc.co.uk/local/northeastwales/hi/people_and_places/history/newsid_8077000/8077550.stm (accessed 29/11/2017).

Brooksbank B. W. L. (2007), *London Main Line War Damage*. Capital Transport, Harrow. ISBN 9781854143099.

Brown B. (2011), *Three Days in May*. Faber & Faber, London. ISBN 9780571282999.

Building Hitler's Supergun: The Plot to Destroy London, (2015). [TV programme] Channel 4.

Churchill W. S. (1948–49), *The Second World War* (6 volumes). Chiswick Press.

Clark R. W. (1965), *Tizard*. MIT Press, Cambridge, USA.

Corrigan G. (2012), *The Second World War: A Military History*. Atlantic Books, London. ISBN 9781843548959.

Crowther J. G. and Whiddington R. (1947), *Science at War*. HMSO.

Department of Engineering News, University of Cambridge, 19, Spring 2016.

Edgerton D. (2012), *Britain's War Machine: Weapons, Resources and Experts in the Second World War*. Penguin Books Ltd, London. ISBN 9780141026107.

Ellis, G. and Mason, P. (2003), 'The Secret Tunnels of South Heighton' BBC. See http://www.bbc.co.uk/history/ww2peopleswar/stories/36/a1128836.shtml (accessed 23/01/2018).

Ellis, G. and Mason, P. (2004), 'The Secret Tunnels of South Heighton' BBC. See http://www.bbc.co.uk/history/ww2peopleswar/stories/19/a2223019.shtml (accessed 23/01/2018).

Ford R. (2013), *Germany's Secret Weapons of World War II*. Chartwell Books, New York. ISBN 9780785830078.

Gilbert M. (1990), *Prophet of Truth: Winston S. Churchill 1922–1939 (Vol. 5)*. Minerva, London. ISBN 0749391030.

Godwin G. (1946), *Marconi 1939–1945: A War Record*. Chatto & Windus, London.

Goodden H. (2007), *Camouflage and Art: Design for Deception in World War 2*. Unicorn, London. ISBN 9780906290872.

Hartcup G. (1970), *The Challenge of War: Scientific and Engineering Contributions to World War Two*. David and Charles. ISBN 0715347896.

Hartcup G. (2003), *The Effect of Science on the Second World War.* Palgrave Macmillan. ISBN 9781403906434.

Hastings M. (2011), *All Hell Let Loose: The World at War 1939–45.* HarperPress, London. ISBN 9780007338092.

Huang R. (2000), *A Lifetime in Academia.* Hong Kong University Press, Hong Kong. ISBN 9622095186.

Hughes J. (1993), *Port in a Storm,* Merseyside Port Folios.

Humphreys R. (2009), *Sidney Sussex College: A History,* Sidney Sussex College, Cambridge, UK. ISBN 9780956359407.

Jenkins R. (2001), *Churchill.* Macmillan. ISBN 0333782909.

Kennedy P. (2013), *Engineers of Victory.* Allen Lane, London. ISBN 9781846141126.

Kerrigan M. (2011), *World War II Plans that Never Happened 1939–45.* Amber Books, London. ISBN 9781907446641.

Lloyd A. H. (1947), *Metropolitan-Vickers Electrical Co.: Contribution to Victory.* Metropolitan-Vickers Electrical Co.

Lukacs J. (1999), *Five Days in London, May 1940.* Yale University Press. ISBN 0300080301.

Macintyre B. (2014), 'Preserve this monument to Hitler's hubris'. *The Times.*

Marston D. (2005), *The Pacific War.* Osprey Publishing, Oxford. ISBN 9781849083829.

Miscellaneous (25 authors) (1940), *Science in War.* Penguin Special No. S74.

Obituary Notices of Fellows of the Royal Society

Oxford Dictionary of National Biography

Overy R. (2006), *Why the Allies Won.* Random House, London. ISBN 9781845950651.

Overy R. (2010), *1939: Countdown to War.* Penguin Books Ltd, London. ISBN 9780141041308.

Purnell S. (2016), *First Lady: The Life and Wars of Clementine Churchill.* Aurum Press, London. ISBN 9781781313077.

Ray J. (1996), *The Night Blitz: 1940–1941.* Cassell, London. ISBN 030435676X.

Roberts A. (2010), *The Storm of War: A New History of the Second World War.* Penguin Books Ltd, London. ISBN 9780713999709.

Rowlinson F. (1947), *Contribution to Victory: Some of the Special Work of the Metropolitan-Vickers Electrical Company Ltd in the Second World War.* Metropolitan-Vickers Electrical Co., Manchester.

Saunders H. S. (1946), *Ford at War.* Ford Motor Co.

Seymour W. J. (1948), *An Account of Our Stewardship: Being a Record of the Wartime Activities of Vauxhall Motors Ltd.*

Smith C. (2009), *England's Last War Against France: Fighting Vichy 1940–42.* Weidenfeld & Nicolson, London. ISBN 9780297852186.

Stone N. (2013), *World War Two: A Short History.* Allen Lane, London. ISBN 9781846141393.

The Richmond Golf Club, 'Local Rules that went Around the World'. Richmond Golf Club, Surry. See http://www.therichmondgolfclub.com/wartime-rules/ (accessed 29/11/2017).

The Times (2012), *The Times,* p.20 (quoting Scottish Association for Marine Sciences archives).

The Times (2014), 'Emlyn Jones'. *The Times.*

Intelligence and Deception

Bailey R. (2008), *Forgotten Voices of the Secret War: An Inside History of Special Operations in the Second World War.* Ebury/IWM. ISBN 9780091918514.

Barbier M. K. (2007), *D-Day Deception: Operation Fortitude and the Normandy Invasion*. Stackpole Books, Mechanicsburg. ISBN 9780811735346.

Copeland B. J. (2010), *Colossus*. Oxford University Press, Oxford. ISBN 9780199578146.

Delaforce P. (1998), *Churchill's Secret Weapons: the Story of Hobart's Funnies*. Pen & Sword Military, Barnsley.

Goodden H. (2007), *Camouflage and Art*. Unicorn Publishing Group, London. ISBN 9780906290873.

Greenberg J. (2014), *Gordon Welchman: Bletchley Park's Architect of Ultra Intelligence*. Pen & Sword Books Ltd, Barnsley. ISBN 9781848327528.

Herivel J. (2008), *Herivelismus and The German Military Enigma*. M. & M. Baldwin, Worcestershire. ISBN 9780947712464.

Hinsley F. H. and Stripp A. (1993), *Codebreakers: The Inside Story of Bletchley Park*. Oxford University Press, Oxford. ISBN 0192801325.

Hinsley F. H. (1993), *British Intelligence in the Second World War*. HMSO. ISBN 0116309563.

Hodges A. (2012), *Alan Turing: The Enigma*. Vintage, London. ISBN 9780099116417.

Judkins P. E. (1992), 'Enigma'. Paper presented at an IEE meeting on the History of Electrical Engineering, University of Surrey, July 1992.

Macintyre B. (2007), *Agent Zigzag: The True Wartime Story of Eddie Chapman: Lover, Betrayer, Hero, Spy*. Bloomsbury Publishing PLC, London. ISBN 9780747587941.

Macrae, S. (2012), *Winston Churchill's Toyshop: The Inside Story of Military Intelligence*. Amberley Publishing Holdings Limited, Stroud. ISBN 9781445608426.

NSA. (2016), 'Cryptological Hall of Honor: Brigadier John Tiltman'. NSA, Maryland. See https://www.nsa.gov/about/cryptologic-heritage/historical-figures-publications/hall-of-honor/2004/jtiltman.shtml (accessed 29/11/2017).

O' Connor J. J. and Robertson E. F. (2003), 'William Thomas Tutte'. University of St Andrews, St Andrews. See http://www-history.mcs.st-and.ac.uk/Biographies/Tutte.html (accessed 29/11/2017).

Scarlett J. (2014), 'Our colossal debt to the genius of Bletchley'. *The Times*.

Rankin N. (2008), *Churchill's Wizards: The British Genius for Deception 1914–1945*. Faber & Faber, London. ISBN 9780571221950.

Welchman G. (1982), *The Hut Six Story: Breaking the Enigma Codes*. McGraw-Hill. ISBN 0070691800.

Younger D. H. (2012), 'William Thomas Tutte. 14 May 1917 – 2 May 2002'. Royal Society Publishing. See http://rsbm.royalsocietypublishing.org/content/roybiogmem/58/283.full.pdf (accessed 29/11/2017).

Williams A. (2013), *Operation Crossbow: The Untold Story of Photographic Intelligence and the Search for Hitler's V Weapons*. Preface Publishing, London. ISBN 9781848093072.

Inter-service Projects

Breuer W. B. (2000), *Secret Weapons of World War II*. John Wiley, New York. ISBN 9780471202127.

Delaforce P. (2005), *Smashing the Atlantic Wall*. Pen & Sword Books Ltd, Barnsley. ISBN 1884152561.

Ford B. J. (2011), *Secret Weapons*. Osprey Publishing, Oxford. ISBN 9781849083904.

Green A. (2013), 'PLUTO – Lifeblood for D-Day and Beyond'. Newcomen Society lecture, 10 April 2013.

Harris A. (1989), 'The Mulberry Harbours'. *Journal of the Newcomen Society*, 61, pp. 81–98.

Haskew M. E. (2007), *Encyclopaedia of Elite Forces in the Second World War*. Pen & Sword Books Ltd, Barnsley. ISBN 9781844155774.

Lochner R., Faber O., and Penney W. G. (1948), 'Bombardon Floating Breakwater'. *The Civil Engineer in War Volume 2*. Institution of Civil Engineers, London.

Yenne W. (2003), *Secret Weapons of World War II*. Berkley Books, New York. ISBN 0425189929.

RN and the War at Sea

Allen J. S. (1983), 'Some Aspects of Technological Change – 1900 to 1939: A Symposium'. *Journal of the Newcomen Society* (now *International Journal for the History of Engineering and Technology*), 55, pp. 31–66.

Barnett C. (1991), *Engage the Enemy More Closely: The Royal Navy in the Second World War*. Hodder & Stoughton, London. ISBN 0340339012.

Braid D. H. P. (1984), 'The Armament of Naval Ships in the Nineteenth Century'. *Journal of the Newcomen Society* (now *International Journal for the History of Engineering and Technology*), 56, pp. 111–132.

British Machine Tool Engineering XXVII, 142, (January – June 1945), Engineering in the Royal Navy.

Broome J. E. (1955), *Make a Signal*. Putnam, London. ISBN 0951448013.

Brown D. K. (1983), *A Century of Naval Construction: The History of the Royal Corps of Naval Constructors*. Conway Maritime Press Ltd, London. ISBN 085177282X.

Brown D. K. (2012), *Nelson to Vanguard: Warship Design and Development 1923–1945*. Seaforth Publishing, Barnsley. ISBN 9781848321496.

Budiansky S. (2013), *Blackett's War: The Men who Defeated the U-Boats and Brought Science to the Art of Warfare*. Vintage Books, New York. ISBN 9780307743633.

Burns R. W. (1992), 'Technology & the Battle of the Atlantic'. Paper presented at an IEE meeting on the History of Electrical Engineering, University of Surrey, July 1992.

Colby R. F. D. (1989), 'The Training of Marine Engineer Officers for the Royal Navy', Institute of Marine Engineers, Centenary Year Conference, Paper 25.

Costello J. and Hughes T. (1977), *The Battle of the Atlantic*. Fontana/Collins. ISBN 0006353258.

Dimbleby J. (2015), *The Battle of the Atlantic: How the Allies Won the War*. Viking, London. ISBN 9780241186602.

Durston A. J., 'The Machinery of Warships'. *Proceedings of the Institution of Civil Engineers*, 119(1), 1985, pp. 17–119 (including correspondence and discussion).

Felton M. (2015), *The Sea Devils: Operation Struggle and the Last Great Raid of World War Two*. Icon Books Ltd, Duxford. ISBN 9781848319943.

Forester C. S. (1943), *The Ship*, Joseph, London.

Hartley A. C. (1948), 'Operation PLUTO'. *The Civil Engineer in War Volume 3*. Institution of Civil Engineers, London.

Honourable Society of Cymmrodorion (2001), *Dictionary of Welsh Biography: 1941–70*. Honourable Society of Cymmrodorion, London. ISBN 0954162609.

Howarth D. (2000), *The Shetland Bus*. Shetland Times Ltd, Lerwick. ISBN 1898852421.

Johns R. F. (1989), 'Technician Training in the Royal Navy – Past, Present and Future'. Institute of Marine Engineers, Centenary Year Conference, Paper 26.

Jolly R. (1989), *Jackspeak: The Pusser's Rum Guide to Royal Navy Slang*. Palamanando Publishing, Torpoint. ISBN 0951430505.

Kemp P. (ed.), (1976), *The Oxford Companion to Ships and the Sea*, Oxford University Press, Oxford. ISBN 0192115537.

Kennedy L. (1974), *Pursuit: The Sinking of the Bismarck*, Collins Fontana. ISBN 0006340148.

Le Bailly L. (1990), *The Man Around the Engine: Life Below the Waterline*. Kenneth Mason Publications Ltd, Hampshire. ISBN 0859373541.

Nesbit R. C. (2012), *The Strike Wings: Special Anti-Shipping Squadrons 1942–45*, Pen & Sword Aviation, Barnsley. ISBN 9781781590287.

Newby G. A. (1992), 'Behind the Fire Doors: Fox's Corrugated Furnace 1877 and the 'High Pressure' Steamship'. *Journal of the Newcomen Society* (now *International Journal for the History of Engineering and Technology*), 64, pp. 143–166.

Offley E. (2011), *Turning the Tide: How a Small Band of Allied Sailors Defeated the U-Boats and Won the Battle of the Atlantic*. Basic Books, New York. ISBN 9780465013975.

Owen D. (2007), *Anti-Submarine Warfare: An Illustrated History*. Pen & Sword Books Ltd, Barnsley. ISBN 9781844157037.

Parsons N. C. (1984), 'Sir Charles Parsons: A Symposium to Commemorate the Centenary of his Invention of the Steam Turbine and Electric Generator – The Origins of the Steam Turbine and the Importance of 1884'. *Journal of the Newcomen Society* (now *International Journal for the History of Engineering and Technology*), 56, pp. 21–58.

Pawle G. (1972), *The Secret War 1939–45*. White Lion, London. ISBN 0856171204.

Penn G. (1984), *HMS* Thunderer: *The Story of the Royal Naval Engineering College, Keyham and Manadon*. Kenneth Mason Publications Ltd, Hampshire. ISBN 0859373215.

Shield R. N. (1989), 'The Impact of Market Forces on Ship Upkeep Strategy (The Past – Historical)'. Institute of Marine Engineers, Centenary Year Conference, Paper 23.

Smith E. C. (1921), 'The Centenary of Naval Engineering – A Review Of The Early History Of Our Steam Navy'. *Journal of the Newcomen Society*, 2, pp. 88–114.

Smith E. C. (2013), *A Short History of Naval and Marine Engineering*. Cambridge University Press, Cambridge. ISBN 9781107672932.

Thomas K. H. W. (1983), *The Royal Corps of Naval Constructors: A Centenary Review*. Royal Institution of Naval Architects.

Varey P. (2012), *Life on the Edge: Peter Danckwerts GC MBE FRS*. PFV Publications, London. ISBN 9780953844029.

Waddington C. H. (1947) *Operational Research in World War II – Operational Research against the U-Boat*, HMSO.

Warner O. (1972), *Great Naval Battles*, Hamlyn, London. ISBN 0600338509.

Watkins G. M. (1974), 'Large Marine Steam Engines'. *Journal of the Newcomen Society* (now *International Journal for the History of Engineering and Technology*), 47, pp.133–148.

Watts A. (1976), *The U-boat Hunters*, Macdonald & Jane's, London. ISBN 035608244X.

Welbourn D. and Crichton T. (2008), 'The Schnorchel: A Short-Lived Engineering Solution to Scientific Developments'. Transactions of the Newcomen Society, 78(2), pp. 213–315.

Winton J. (1998), *Cunningham: The Greatest Admiral Since Nelson*. John Murray Publishers Ltd, London. ISBN 0719557658.

Army and the War on Land

Bailey D. C. and Carey W. M. D. (1948), 'The Bailey Suspension Bridge'. *The Civil Engineer in War Volume 1*. Institution of Civil Engineers, London.

Bailey D. C., Foulkes R. A., and Digby-Smith R. (1948), 'The Bailey Bridge and its Developments'. *The Civil Engineer in War Volume 1*. Institution of Civil Engineers, London.

Begbie D. L. G. and Roberts G. (2014), 'Bridging in the Second World War: An Imperative to Victory'. *Proceedings of the Institution of Civil Engineers*, [online] 167, EH2 p. 111. See org/10.1680/ehah.13.00022 (accessed 29/11/2017).

Connolly T. W. C. (1856), *History of the Royal Sappers and Miners*. Longmans, London.

Cooper M. D. (2006), *A Short History of the Corps of Royal Engineers*, Institution of Royal Engineers, Chatham. ISBN 0903530287.

Daily Telegraph (2011), 'John Stone'. *Daily Telegraph.*

Harpur B. (1991), *A Bridge to Victory: The Untold Story of the Bailey Bridge.* HMSO. ISBN 0117726508.

Hogg O. F. G. (1963), *The Royal Arsenal.* Oxford University Press, Oxford.

Hudson P. G. (1948), 'The Development and Construction of Airfields and Runways for the Royal Air Force, 1939–1945'. *The Civil Engineer in War Volume 1.* Institution of Civil Engineers, London.

Hull G. B. G. (1948), 'A Short Description of the 'Hull' Bridge across the River Shatt-el-Arab, at Margil (Basra), Iraq'. *The Civil Engineer in War Volume 1.* Institution of Civil Engineers, London.

Jellett J. H. (1948), 'The Lay-out, Assembly, and Behaviour of the Breakwaters at Arromanches Harbour (Mulberry 'B')'. *The Civil Engineer in War Volume 2.* Institution of Civil Engineers, London.

Joiner J. H. (2001), *One More River to Cross.* Pen & Sword Books Ltd, Barnsley. ISBN 0850527880.

Joiner J. H. (2011), 'The Story of the Bailey Bridge'. *Engineering History and Heritage,* 164, EH2, p. 65–72.

Kennett B. B. and Tatman J. A. (1970), *Craftsmen of the Army.* Corps of the Royal Mechanical and Electrical Engineers.

Lochner R., Faber O., and Penney W. G. (1948), 'The 'Bombardon' Floating Breakwater'. *The Civil Engineer in War Volume 2.* Institution of Civil Engineers, London.

Nalder R. F. H. (1953), *The History of the British Army's Signals in the Second World War.* Royal Signals Institution.

Napier G. (1998), *The Sapper VCs.* HMSO. ISBN 0117728357.

Napier G. (2005), *Follow the Sapper.* Institution of Royal Engineers, Chatham. ISBN 0903530260.

Pakenham-Walsh R. P. (1993), *History of the Corps of Royal Engineers (1938–1948) Volumes VIII.* Institution of Royal Engineers, Chatham.

Pakenham-Walsh R. P. (1993), *History of the Corps of Royal Engineers (1938–1948) Volumes IX.* Institution of Royal Engineers, Chatham.

Pam D. (1998), *The Royal Small Arms Factory, Enfield and its Workers.* David Pam, Enfield. ISBN 0953227103.

Porter W. (1889), *History of the Corps of Royal Engineers.* Longmans.

Various, *History of the Corps of Royal Engineers.*

Ware P. (2012), *Special Forces Vehicles: 1940 to the Present Day.* Pen & Sword Military, Barnsley. ISBN 9781848846425.

Younger T. (2004), *Blowing our Bridges: A Memoir from Dunkirk to Korea via Normandy.* Pen & Sword Books Ltd, Barnsley. ISBN 1844150518.

RAF and the War in the Air

Air Ministry. (1963), *The Origins and Development of Operational Research in the Royal Air Force.* London.

Bond S. and Forder R. (2011), *Special Ops Liberators: 223 (Bomber Support) Squadron, 100 Group, and the Electronic War.* Grub Street, London. ISBN 9781908117144.

Brooks R. J. (1993), *Sussex Airfields in the Second World War.* Countryside Books, Newbury. ISBN 1853062596.

Burns R. W. (1995), 'Technology and Air Defence, 1914 to 1945', *Journal of the Newcomen Society,* 67/1.

Burns R. W. (1996), The Problem of Detecting Hostile Aircraft at Night (1935–1941). IET.

Burns R. W. (1996), 'The Problem of Detecting Hostile Aircraft at Night'. Paper presented at an IEE meeting on the History of Electrical Engineering, University of York, July 1996.

Burns R. W. (1997), 'The Development of Methods of Detecting Hostile Aircraft at Night: 1939–41'. *International Journal for the History of Engineering & Technology,* 69A, pp. 1–21.

Coles D. and Sherrard P. (2012), *The Four Geniuses of The Battle of Britain: Watson-Watt, Henry Royce, Sydney Camm and RJ Mitchell.* Pen & Sword Aviation, Barnsley. ISBN 9781848847590.

Cross G. E. (2001), *Jonah's Feet are Dry: The Experience of the 353rd Fighter Group During World War II.* Thunderbolt, Ipswich. ISBN 0954116402.

Daily Telegraph. (1999), 'Ronnie Harker'. *Daily Telegraph.*

Davies B. (2013), *617 Squadron The Dam Busters 70th Anniversary,* Newsdesk Media and RAF Museum. ISBN 9781906940713.

Davis J. (2009), *Britain's Forgotten Heroes.* Pearl Press. ISBN 9780956307651.

East Grinstead Museum, 'Guinea Pig Club – Rebuilding Bodies and Souls'. East Grinstead Museum, East Grinstead. See https://www.eastgrinsteadmuseum.org.uk/guinea-pig-club/ (accessed 29/11/2017).

Harvey A. D. (2015), 'German Aircraft Design during the Third Reich'. *International Journal for the History of Engineering and Technology,* 85 (2).

Jones R. V. (1978), *Most Secret War.* Book Club Associates, London. ISBN 9780241897461.

Judkins P. (2012), 'Making Vision into Power: Power Struggles & Personality Clashes in the Development of Allied Radar1941–45'. *International Journal for the History of Engineering and Technology,* 82(1), pp. 82–124

Perera L. (2011), *Spitfire.* Instinctive Product Development. ISBN 9781907657931.

Terraine J. (2010), *The Right of the Line: The Role of the RAF in World War II.* Pen & Sword Books Ltd, Barnsley. ISBN 9781848841925.